Biology and Chemistry of Plant Trichomes

Edited by
Eloy Rodriguez
Patrick L. Healey
and
Indira Mehta
University of California
Irvine, California

Plenum Press • *New York and London*

Library of Congress Cataloging in Publication Data

Main entry under title:

Biology and chemistry of plant trichomes.

"Proceedings of a symposium on biology and chemistry of plant trichomes, held July 11-16, 1980, at the joint meeting of the Botanical Societies of America and Canada, in Vancouver, Canada"—T.p. verso.

Includes bibliographies references and index.

1. Trichomes—Congresses. I. Rodriguez, Eloy. II. Healey, Patrick L. III. Mehta, Indira. IV. Botanical Society of America.

QK650.B56 1983   581..4'7   83-11150

ISBN 0-306-41393-0

Proceedings of a symposium on Biology and Chemistry of Plant Trichomes, held July 11-16, 1980, at the joint meeting of the Botanical Societies of America and Canada, in Vancouver, Canada

©1984 Plenum Press, New York
A Division of Plenum Publishing Corporation
233 Spring Street, New York, N.Y. 10013

All rights reserved

No part of this book may be reproduced, stored in a retrieval system, or transmitted in any form or by any means, electronic, mechanical, photocopying, microfilming, recording, or otherwise, without written permission from the Publisher

Printed in the United States of America

PREFACE

For centuries it has been recognized that plants elaborate trichomes (hairs) that produce natural chemicals useful to the plant and man. These trichomes consist of one or more cells which are derived from single protodermal cells and have a variety of functions. They defer phytophagous insects and function as protection from excess temperature drop or water loss by covering the plant surface with a trapped air space. Glandular trichomes synthesize, metabolize, or accumulate and secrete terpenoids, phenolics, mucoproteins, and resins. Stinging hairs of nettles and other plants not only deter herbivory but in many cases elicit severe cases of skin dermatitis.

Although a number of reviews have been published in the last 15 years, few have attempted to cover the biology and chemistry of plant trichomes. With this in mind, a symposium was organized to bring together scientists working on diverse aspects of plant trichomes. The symposium, on which this volume is based, was presented at the joint meeting of the Botanical Society of America and Canadian Botanical Society held in Vancouver, Canada on July 11-16, 1980. Speakers were chosen because of their pre-eminence in specific aspects of plant trichome research. The symposium on plant trichomes focused on these areas: development (Peterson and Vermeer), ultrastructure (Mahlberg et al. and Behnke), physiology (Thomson and Healey), chemistry (Croteau & Johnson and Kelsey, Reynolds & Rodriguez), systematic use (Wollenweber), and ecological significance (Ehleringer).

The symposium was sponsored by the Developmental, Phytochemical, and Structural Sections of the Botanical Society of American. We also acknowledge the University of British Columbia for hosting the symposium and the National Science Foundation (# PCM-8007018) for financial support.

The editors also wish to acknowledge the input provided by Dr. Tom J. Mabry (University of Texas, Austin) who not only reviewed many of the manuscripts but also provided the inspiration for the Symposium. We are also thankful to Ms. Elena Viramontes (UCI) for her assistance in editing and the Word Processing Staff in Developmental and Cell Biology for typing the manuscripts.

CONTENTS

1. Plant Trichomes – Structure and Ultrastructure: General Terminology, Taxonomic Applications, and Aspects of Trichome-Bacteria Interaction in Leaf Tips of *Dioscorea*
   H.-Dietmar Behnke ... 1

2. Structure, Development and Composition of Glandular Trichomes of *Cannabis sativa* L.
   Paul G. Mahlberg, Charles T. Hammond, Jocelyn C. Turner & John K. Hemphill ... 23

3. The Systematic Implication of Flavonoids Secreted by Plants
   Eckhard Wollenweber ... 53

4. Histochemistry of Trichomes
   Robert L. Peterson & Janet Vermeer ... 71

5. Cellular Basis of Trichome Secretion
   William W. Thomson & Patrick L. Healey ... 95

6. Ecology and Ecophysiology of Leaf Pubescence in North American Desert Plants
   James Ehleringer ... 113

7. Biosynthesis of Terpenoids in Glandular Trichomes
   Rodney Croteau & Mark A. Johnson ... 133

8. The Chemistry of Biologically Active Constituents Secreted and Stored in Plant Glandular Trichomes
   Rick G. Kelsey, Gary W. Reynolds & Eloy Rodriguez ... 187

   Participants ... 243

   Index ... 245

PLANT TRICHOMES - STRUCTURE AND ULTRASTRUCTURE:

GENERAL TERMINOLOGY, TAXONOMIC APPLICATIONS, AND ASPECTS OF

TRICHOME-BACTERIA INTERACTION IN LEAF TIPS OF *DIOSCOREA*

H.-Dietmar Behnke

Zellenlehre, Universität Heidelberg, Im Neuenheimer Feld 230

D-6900 Heidelberg, Fed. Rep. of Germany

ABSTRACT

Plant trichomes (epidermal appendages as a whole) comprise such structural and functional extremes as glandular hairs, trigger hairs, absorbing root hairs, scales, papillae, and micro-hairs. Examples for these and for a more descriptive classification of trichomes into uni- and multicellular, branched and unbranched, uni- and multiseriate, stellate, peltate, T-shaped, dendritic, and other forms are given. The entire indumentum of a plant and the degree of pubescence contribute to its individual and species-specific characters. Single trichome types, however, have in many cases been successfully applied to the classification of different taxa: results with Leguminosae, Cruciferae, and Urticales, taken from the literature, are reviewed. Ultrastructural and functional aspects of two trichome types of Dioscorea macroura *investigated by the author are presented. Stalked glandular hairs, situated within furrows formed by the leaf tips, secrete acid polysaccharides which attract bacteria. A massive appearance of bacteria, on the other hand, initiates protuberances of the epidermal cells which develop into uniseriate multicellular trichomes, line all the furrows harboring bacteria, and may contribute to a disintegration of bacteria.*

1. INTRODUCTION

Plant trichomes were among the first anatomical features recognized and depicted by the early microscopists of the 17th century: Hooke (1665), Grew (1682), and Malpighi (1686). Since then, trichomes have become one of the more intensively studied plant characters, mainly due to their ubiquity among higher plants, their different functions and the extreme morphological variations of both the single trichome and the composition of the entire indumentum.

Plant trichomes comprise such structural and functional extremes as hairs, glands (i.e. glandular hairs, capitate hairs, colleters), scales (or peltate hairs) and sometimes, emergences and papillae. In general, the definition of the trichome as an epidermal appendage would exclude emergences since they derive from epidermal and subepidermal layers and probably would also exclude papillae, i.e. protrusions of the periclinal outer epidermal cell walls. Both formations, however, intergrade with real trichomes and without knowledge of their development, it is often not possible to make a clear distinction. Papillae, e.g., can be easily confused with unicellular trichomes and probably should be regarded as one of their subgroups. In the absence of a clear definition, the distinction between the two is made by measuring the quotient between the width and height of an epidermal appendage (Barthlott and Ehler 1977).

Monographic descriptions of the entire indumentum and of specialized hair types, including functional, developmental and taxonomic aspects, have been undertaken for a great variety of plant taxa, from the species up to the family level. Extensive reviews are by Netolitzki (1932) and Uphoff (1962).

1.1. *Trichome Terminology*

Despite the large volume of previous work (or probably because of it, and certainly to some extent due to its wide historical background), so far no satisfactory and generally accepted morphological classification of the various trichome types has been produced, although several attempts have been made in the past. A logical and detailed classification, however, is needed for taxonomic purposes, which represents one of the major applications of trichome data.

Most authors who attempt to classify the entire wealth of trichome forms use a mixture of anatomical and morphological criteria for their distinctions. Some frequent distinctions made are between unicellular and multicellular, simple and compound, glandular and nonglandular trichomes. Dickison (1974), for example, adopting mainly suggestions made from the genus *Solanum* by Roe (1971), discerns three different types of trichomes:

simple, complex and glandular, each of which is further divided
into different sections with specialized forms. Among the simple
(i.e. unbranched) trichomes, Dickison (1974) treats unicellular as
well as multicellular trichomes. The multicellular trichomes in
turn, may be uni- or multiseriate. The complex type refers to
branched trichomes. The different forms of trichomes discerned
in each of these sections, as well as those distinguished by other
authors, are described rather incompletely using a trivial
terminology accompanied by simple line drawings. As a consequence, there is almost an endless vocabulary in use in the
trichome literature with additions arising from almost every new
investigation. Payne (1978) compiled a glossary of more than 300
terms derived from only 16 major reference sources! To add to
the confusion, Payne also records some 150 terms strictly referring to the characteristics of the entire indumentum, the description of which for some studies (e.g., taxonomic) is of
equal or even of more importance than that of the single trichome.
Therefore, beyond general classifications, it seems hopeless to
bring other than alphabetical order to the great number of terms.

A new start towards clarifying and unifying the description
of trichome morphology and anatomy as well as methods of trichome
analysis has been made by Theobald et al. (1980) as part of an
introductory volume of the second edition of Metcalfe and Chalk's
"Anatomy of the Diotyledons." Their concept of an adequate
trichome description includes a four step examination of the
(1) indumentum, i.e. overall surface appearance; (2) morphology of
the individual trichome; (3) trichome complement, i.e. the coexistence of different trichome types on a given surface (of
either a specific organ or taxon) and (4) histology (anatomy) of
individual trichomes. Special emphasis is placed on a clear
distinction between morphological and anatomical characters
(steps 2 and 4) which, in earlier descriptions, were often
intermingled.

For use in the purely morphological section (2) Theobald
et al. (1980) distinguishes seven trichome types, each of which
has several subtypes: (a) papillae, (b) simple (unbranched)
trichomes, (c) two- to five-armed trichomes, (d) stellate
trichomes, (e) scales, (f) dendritic (branching) trichomes, and
(g) trichomes of specialized types.

Anatomical details facilitate the comparison of trichomes
and taxon-specific surfaces. Features like glandular/nonglandular, uni/multicellular, uni/multiseriate, surface
sculpturing (warts, striations, etc.), different cell types
within the trichome, wall impregnations and others are listed
among step 4 (see Theobald et al. 1980, for a complete sequence).

A general acceptance of this proposal as a guideline to the
study of any trichome would aid in the simplification of surface
comparisons for the use of taxonomists and systematists.

Fig. 1. *Arabidopsis sp* (Brassicaceae): Dendritic unicellular trichome with micropapillate surface sculpturing from the lower surface of a leaf. x 250.
(SEM micrograph by Dr. W. Barthlott)

1.2. *Trichome Ultrastructure in Relation to Taxonomy*

After Solereder's 1899 work appeared, the great value of trichomes for the characterization of plant taxa could no longer be neglected. Their taxonomic and phylogenetic implications are now considered in all modern basis texts in plant systematics. Staudermann (1924), Solereder and Mayer (1928), Metcalfe and Chalk (1950), Metcalfe (1960, 1971) and Hummel and Staesche (1962) continued the taxonomic application of trichomes, extended it to all angiosperms and eventually included aspects of trichome ultrastructure (Metcalfe and Chalk, 1980).

The three examples used in this account can only roughly indicate the great potential of the application of trichome ultrastructure to different levels of angiosperm taxonomy.

1. The Brassicaceae are a prominent example of a higher taxon in which different classifications have considered trichome characters at almost every level (carried to its utmost extreme by Prantl, 1891). Among the different trichomes which make up

the indumentum of individual members of the family Brassicaceae, the most widespread type, used as a character in familial assignment, is a dendritic trichome, unicellular and covered with cuticular micropapillae (Rollins and Banerjee 1975, 1976; Barthlott 1980). Its morphology can be described as stalked, few to many branched, and terminally branching. The example chosen in Fig. 1 shows one of the few-branched forms and very prominently demonstrates the characteristic micropapillate surface features, derived from a cuticular folding pattern (Barthlott, unpublished).

Rollins and Banerjee (1975) chose the genus *Lesquerella* as an example demonstrating that certain trends in the specialization of a single cell can be used to retrace the phylogeny of a taxon. In their words, the trichome diversity of *Lesquerella* "may be viewed as an exemplar of the diversification of a single cell type produced by evolutionary selection over time" (p. 3).

The following are the major features of trichome diversification in *Lesquerella* which have been used by Rollins and Banerjee (1976) to demonstrate species affinities: (a) transition from dendritic to stellate trichomes; (b) increase in branching complexity; (c) occurrence and amount of ray fusion; (d) increase of webbing between rays, and (e) presence or absence of an umbo, i.e. a convex elevation at the center of the trichome. Transitional forms between dendritic trichomes with irregularly arising, unequal branches and stellate trichomes with equal branches radiating from a center have a fanlike arrangement of the branches, leaving a gap at the central attachment point. Branching complexity refers to the forking (one or two times) of the rays at a distance from the center. Ray fusion sometimes results in a scalelike trichome and is to be distinguished from webbed trichomes where the webbing material is very distinct from the ray proper.

The characteristic surface sculpturing named "tuberculate" by Rollins and Banerjee (1975, 1976) and identified by Barthlott (1980) as cuticular micropapillae is present throughout almost all species, although some of the few-branched dendritic trichomes may have a smooth surface.

2. The subtribal classification of the tribe *Phaseoleae* (Fabaceae) is largely supported by the presence of distinct trichome types which, together with some other leaf characters, were studied by Lackey (1978). While nonglandular uniseriate simple trichomes and glandular trichomes of various types are common in the entire tribe, long simple trichomes with a multicellular (glandular?) complex on top of a single basic cell, and with a terminal biseriate part ("bulbous based hairs") as well as glandular, multicellular sessile scales ("vesicular glands"), are characteristically restricted to the subtribe Cajaninae (see also Lackey 1979). All species of *Erythrina*, and of no other taxon in the tribe, contain multicellular two-armed (T-shaped, Y-shaped) or many-branched dendritic trichomes. Probably the most peculiar hair type is the "hooked hair," a

short two-cellular simple trichome (Figs. 2, 3) mainly present in *Phaseolus s. l.* and *Clitoria - Periandra - Clitoriopsis - Centrosema*.

3. Probably the most striking coincidence in the trichome ultrastructure of two different taxa was revealed by Barthlott and Ehler (1977) when they compared the upper leaf surfaces of *Cannabis sativa* and *Humulus lupulus* (both from the family Cannabaceae). Both species are characterized by short unicellular simple trichomes, the walls of which are impregnated with silica and covered with cuticular micropapillae (Figs. 4, 5). Actually, the micropapillae are a continuation of the cuticular folding present on the surface of the surrounding epidermal cells. These features taken together are so unique that they would directly indicate a very close relationship between the two genera, even if the rest of the similarities between them were not known. An inspection of other families of the Urticales demonstrated this peculiar trichome type to be also present in Urticaceae, Cecropiaceae, and Moraceae (Bathlott 1981, Hardin 1981).

1.3. *Trichomes in the Leaf Tips of* Dioscorea macroura: *Development, Ultrastructure and Interaction with Bacteria*

Leaf tips of many plants develop well ahead of other parts of the lamina. A taxon which shows this phenomenon in almost all its members is the monocotyledonous family Dioscoreaceae. In this family, the West African species *Dioscorea macroura* Harms, however, stands out as one of the two only species with a thick and fleshy leaf tip (Fig. 6) the other being *D. sansibariensis* (Ayensu 1972). Since the leaf tip is also extremely long, often over five centimeters, this character was incorporated into the species diagnosis of *D. macroura* (Harms, 1897).

The upper surface of the fleshy acumen is ridged, with the ridges leading into inner cavities which originate from infoldings of the lamina. During leaf ontogeny, at first two infoldings are formed, one after another, each of which eventually becomes subdivided into two cavities. From the tip to the base of the acumen the number of cavities gradually increases.

The first detailed description of the anatomy of the leaf tips was given by Orr (1923) who, because of the secretory

Figs. 2, 3. *Phaseolus vulgaris* (Fabaceae): "Hooked hairs" - two-cellular simple trichomes - on transition zone between leaf blade and petiole (2) and on upper leaf surface (3). Arrow points to transverse wall between the two cells. 2: x 250 3: x 1000
(SEM micrographs by Dr. W. Barthlott)

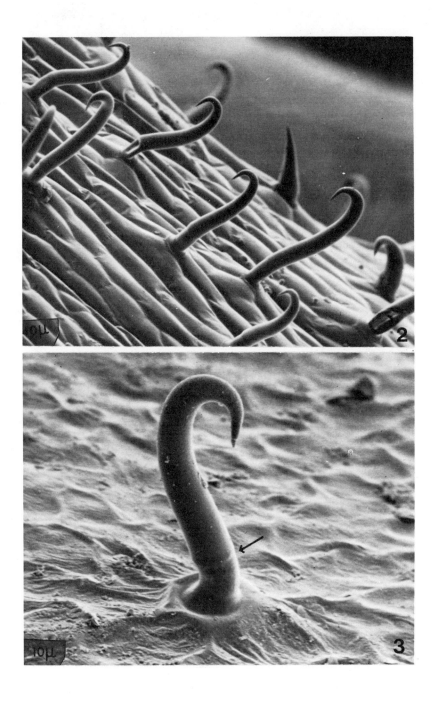

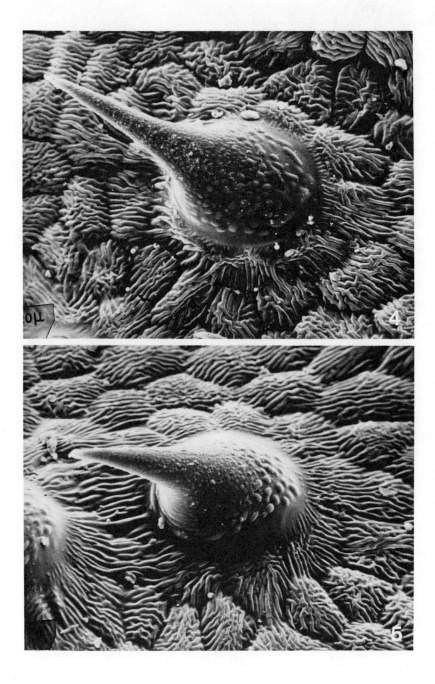

function which was detected, named the cavities "glands". Orr also identified N-fixing and other bacteria in the cavities and on the basis of the high nitrogen content in the acumen suggested a bacterial symbiosis, comparable to those in the Rubiaceae (Lersten and Horner 1976). Schaede (1939) did not detect any bacteria in either the cavities of the very young leaves or at the vegetative cone and therefore questioned the presence of a symbiosis. According to his view, bacteria are first attracted by necrotic glandular trichomes. Such trichomes were mentioned also by Orr (1923) who found them covering the lamina of very young leaves but vanishing later. Schaede's anatomical investigations further proved that the ducts which connect the cavities to the exterior are closed during leaf ontogeny. The trapped bacteria respond to the sealing with rapid reproduction, to which the plant reacts by the formation of simple, long and uniseriate trichomes, filling the entire cavities. According to Schaede (1939), these trichomes contribute to the disintegration of bacteria, since they were not formed if the cavities did not host bacteria.

These suggestions concerning the interactions between the leaf tip cavities of *D. macroura* and their bacterial hosts prompted us to investigate the ultrastructural development of the cavities, the nature of their secretions, and the fate of the enclosed bacteria.

The initiation of the cavities can be visualized by cross-sectioning the tip of the first detectable leaf the lamina of which consists almost exclusively of the acumen. A first cavity is formed by basal extension of a furrow which is completely lined by epidermal cells (Fig. 7). The small cavity encloses a few peltate scales (Fig. 8, arrows) which, when protrayed with the electron microscope, prove to be glandular and multicellular, with a six-celled head connected by a single stalk cell to the basal cell (Figs. 9, 12).

Such glandular scales are present all over the upper surface of young leaves from where, during the process of infolding of

Figs. 4, 5.

Fig. 4. *Cannabis sativus* (Cannabaceae), Fig. 5 *Humulus lupulus* (Cannabaceae): "Lithocysts" - silicified unicellular simple trichomes on upper leaf surfaces. Both trichomes are covered with micropapillae which are the continuation of the cuticular folding present on the surface of the epidermal cells. x 500
(SEM micrographs by Dr. W. Barthlott)

the leaf tips, they eventually get into the cavities (Figs. 10, 11). Once they are located inside the cavities, their peak secretion is already past. Only occasionally, in very young leaf tips, is a still active trichome encountered (Fig. 12). The cytoplasm of their head cells is abundant in dictyosomes which produce Golgi vesicles that migrate to the plasmalemma (Fig. 13). The discharged contents of the vesicles are first visible outside the plasmalemma (Fig. 14: *). They then start to permeate through the cellulose parts of the periclinal cell walls and, while becoming detached from the cutinized layers, accumulate in the subcuticular space (Fig. 14: o; also bent surface of scale in Fig. 9).

By chromatographic analysis (thin-layer chromatography of hot-water extracts of leaf tips which had been hydrolyzed with tri-fluoracetic acid), the secretion was identified as a typical mucopolysaccharide, mainly composed of galactose, arabinose and xylose. Not unlike other plant mucilages (Schnepf 1969; Fahn 1979) these show an increase in stainability (probably dependent on the degree of polymerization) during their passage from the Golgi vesicles to the subcuticular space and finally to the exterior of the scale. This is clearly demonstrated in the micrographs (follow arrows in Figs. 13, 14).

As soon as the mucilage has left the subcuticular space, bacteria (Fig. 15) can be found residing in the furrows of the leaf tips (starting from about the third leaves down to the old ones). At about the same time, the epidermal cells which line the interior part of the cavities resume their mitotic activity and start growing into uniseriate simple trichomes (Fig. 16).

These eventually fill up all the cavities and thus contribute to their glandular appearance, as is well-known from the early light micrographs by Orr (1923) and Schaede (1939). A transverse

Figs. 6-9. *Dioscorea macroura*, leaf tips.

Fig. 6. Fleshy tip of an old leaf (upper left) compared to the tips of young leaves which in earliest leaf pairs make up more than 50% of the blade. 2 x.

Fig. 7. Transverse section through tip of first leaf with first cavity formed. LM: x 130.

Fig. 8. Detail from Fig. 7 showing epidermal cells lining the cavity and a few glandular trichomes at its base (arrows). LM: x 640.

Fig. 9. Glandular trichome with stalk cell and six-celled head from upper surface of a young leaf. Cuticle of head cells blown up in part by secretion products. SEM: x 1400.

PLANT TRICHOMES—STRUCTURE AND ULTRASTRUCTURE 11

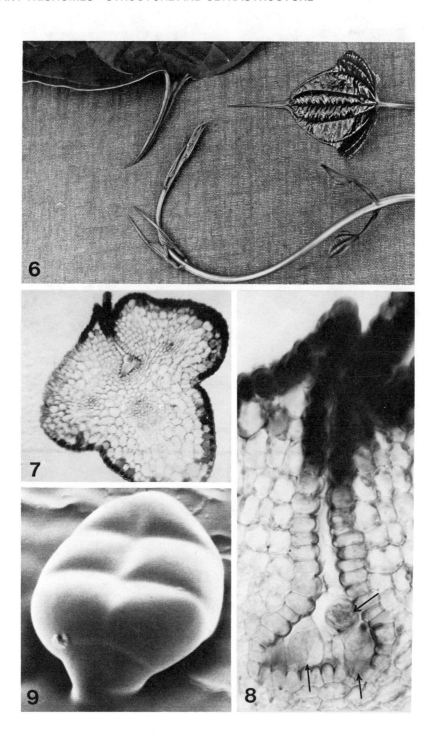

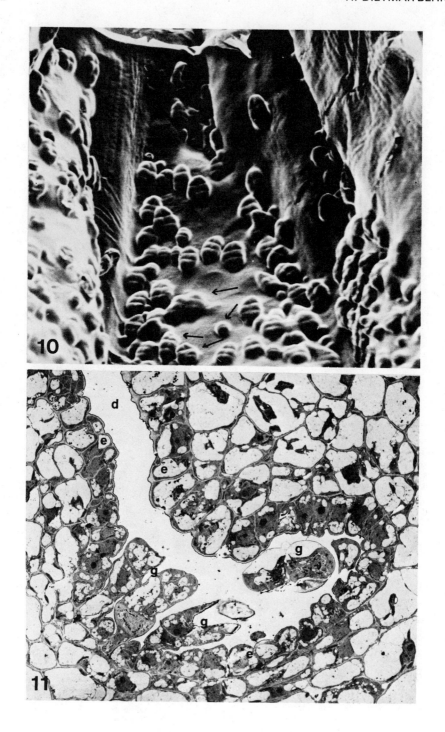

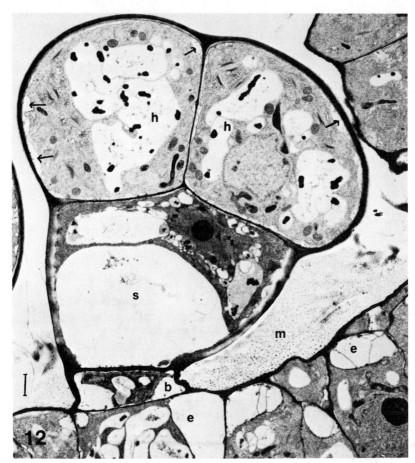

Fig. 12. *Dioscorea macroura*. Glandular trichome from inner cavity of a transverse sectioned tip of first leaf. Head cells (h) are still actively secreting (arrows point to secretion products accumulated outside plasmalemma). s = stalk cell; b = base cell (cut obliquely: not showing actual location inside epidermis). e = epidermal cells lining cavity. m = mucopolysaccharide secreted from head cells. TEM: x 5000, marker = 1 µm.

Fig. 10. *Discorea macroura*. Surface view of a young leaf tip showing infolding of blade with inner cavity in process of formation (top). Epidermis studded with glandular trichomes, in part covered by their secretion products (arrows). SEM: x 140.

Fig. 11. *Discorea macroura*. Transverse section through a tip of a second leaf showing inner cavity and duct (d) still open to the exterior, both lined by epidermal cells (e). Inside the cavity are a few glandular trichomes (g). TEM: x 1000.

Figs. 13-15. *Dioscorea macroura*.

Fig. 13. Part of a head cell from a glandular trichome within a tip of a first leaf: Active dictyosomes (D) give rise to golgi vesicles, the contents of which accumulate outside the plasmalemma (*). TEM: x 12,000.

Fig. 14. Parts of three head cells from glandular trichome within a tip-cavity of a second leaf: The mucopolysaccharides secreted by dictyosomes (D) first accumulate outside the plasmalemma (*), then penetrate the cellulosic wall (W) and stay (o) between wall and cuticle (C), before finally they are released into the tip-cavity. TEM: x 8000.

Fig. 15. Bacteria inside a tip-cavity of a fourth leaf and located among the mucopolysaccharide (m). TEM: x 20,000, marker = 1 µm.

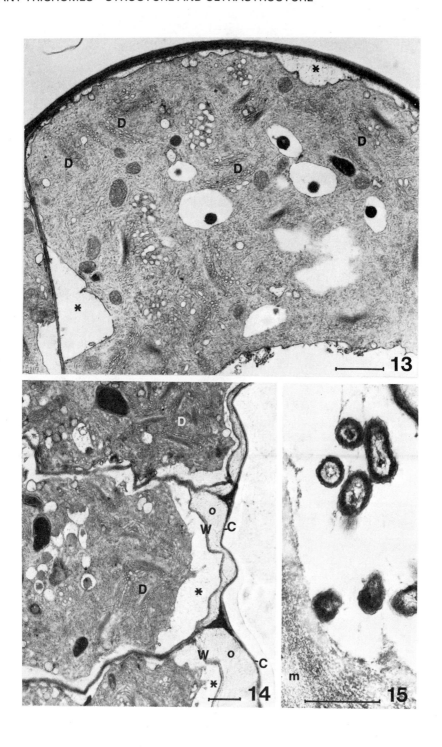

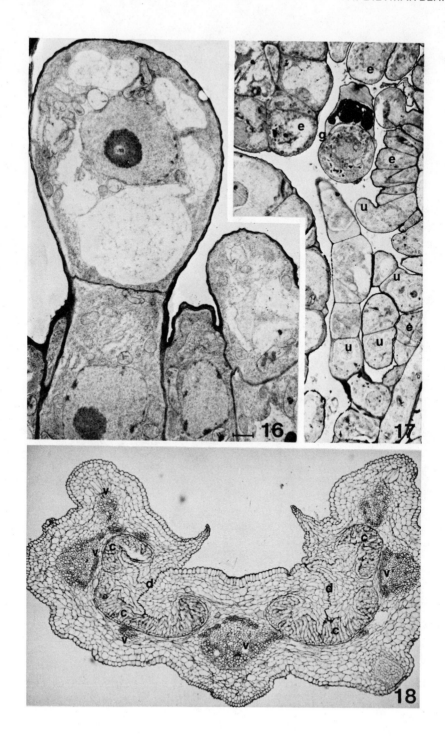

Figs. 16-18. *Dioscorea macroura*.

Fig. 16. Outgrowths of epidermal cells lining the tip-cavities in process of formation of uniseriate multicellular trichomes. TEM: x 5000, marker = 1 µm.
Fig. 17. Transverse section through a tip-cavity of a third leaf. Some of the epidermal cells (e) have already started growth into uniseriate trichomes (u). Inside the cavity are also parts of a degenerating glandular trichome (g). TEM: x 1700.
Fig. 18. Transverse section through the tip of a tenth leaf. Four internal cavities (c) visible; all are filled with uniseriate trichomes. The ducts (d) which formerly connected the cavities to the exterior are now occluded. The cavities are surrounded by vascular bundles (v) supplying them with water and assimilates. LM: x 65.

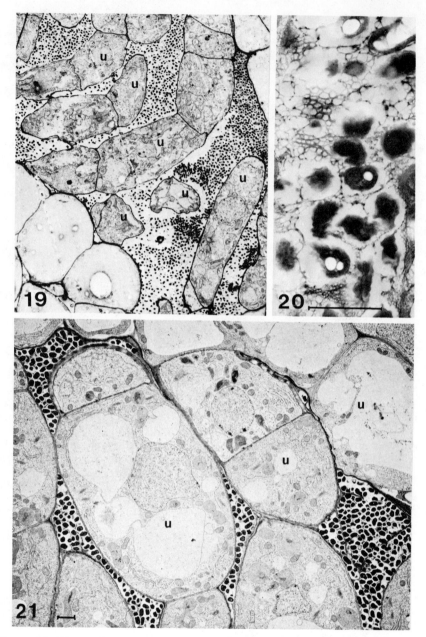

Figs. 19–21. *Dioscorea macroura*. Transverse sections through tip cavities of older leaves. Uniseriate trichomes (u) take over most of the space of the cavities (19), until only small areas are left (21) which are completely occupied by bacteria. The bacteria which still remain among the remnants of the mucopolysaccharide, are eventually degraded (20). All TEM. 19: x 1200, 20: x 20,000, 21: x 4000, marker = 1 μm.

section through a tip of an old leaf viewed with the LM (Fig. 18) illustrates the tight packing of the trichomes, while thin-sectioning for TEM of progressively older leaf tips demonstrates how the cavities are filled. The uniseriate trichomes bending into the lateral extensions of the cavities grow till there is no more space left (Figs. 16, 17, 19). All gaps in between the trichomes are invaded by the rapidly propagating bacteria (Fig. 21). Finally, in old leaves, the bacteria seem to disintegrate (Fig. 20).

The evidence from ultrastructural and chromatographic studies accumulated so far seems to support Schaede's (1939) arguments that the leaf tip bacteria are parasitic rather than symbiotic. Certainly, it is the mucopolysaccharide produced by the glandular scales that attracts the bacteria and is the substrate upon which they feed. The formation of the uniseriate hairs, however, which occurs at about the time when the bacteria start their bulk reproduction and after occlusion of the ducts is indicative of a closer interaction which will hopefully be elucidated in future studies.

ACKNOWLEDGEMENTS

The present paper was read during the Botany 80 symposium at Vancouver by Lise Bolt Jørgenson (København). W. Barthlott (Heidelberg) kindly gave permission to incorporate unpublished results of his trichome studies (as Figs. 1-5) and took the SEM micrographs of *Dioscorea*. Cand. biol. W. Siller started a detailed analysis of the ontogeny and composition of the leaf tips of *Dioscorea*, some results of which were included herein. Mrs. L. Pop, Mrs. D. Laupp, and Miss C. Walter (Heidelberg) provided skillful assistance. I express my sincere thanks to all of them. Supported in part by Deutsche Forschungs-gemeinschaft.

2. REFERENCES

Ayensu, E.S., 1972, Dioscoreales, *In:* C.R. Metcalfe (ed.), Anatomy of the Monocotyledons VI. Clarendon Press, Oxford.
Barthlott, W., 1980, Morphogenese und Mikromorphologie komplexer Cuticular-Faltungsmuster an Blüten-Trichomen von *Antirrhinum* L. (Scrophulariaceae). Ber. Dtsch. Bot. Ges. 93:379-390.
Barthlott, W., 1981, Epidermal and seed surface characters in plants: Systematic applicability and some evolutionary aspects. Nord. J. Bot. 1:345-355.
Barthlott, W. and N. Ehler, 1977, Rasterelektronenmikroskopie der Epidermis-Oberflächen von Spermatophyten. Trop. Subtrop. Pflanzenwelt 19:367-467.

Dickison, W.C., 1974, Trichomes. *In:* A.E. Radford, W.C. Dickison, J.R. Massey, and C.R. Bell (eds.), Vascular Plant Systematics, pp. 198-202. Harper and Row, New York.
Fahn, A., 1979, Secretory tissues in plants. Academic Press, London, New York and San Francisco.
Grew, N., 1682, The anatomy of plants with an idea of a philosophical history of plants. London.
Hardin, J.W., 1981, Atlas of foliar surface features in woody plants. II. *Broussonetia, Morus,* and *Maclura* of North America. Bull Torr. Bot. Club 108:338-346.
Harms, H.A.T., 1897, *Dioscorea macroura* Harms n.sp. Notizbl. Kgl. Bot. Gart. u. Mus. Berlin 1:266-267.
Hooke, R., 1665, Micrographia or some physiological description of minute bodies made by magnifying glasses with observations and inquiries thereupon. London.
Hummel, K. and K. Staesche, 1962, Die Verbreitung der Haartypen in den natürlichen Verwandtschaftsgruppen. *In:* W. Zimmermann and P.G. Ozenda (eds.). Encycl. Plant Anatomy IV, 5:209-250. Gebr. Bornträger, Berlin-Nikolassee.
Lackey, J.A., 1978, Leaflet anatomy of Phaseoleae (Leguminosae: Papilionoideae) and its relation to taxonomy. Bot. Gaz. 139:436-446.
Lackey, 1979, Leaflet hairs of *Adenochilos* (Leguminosae - Papilionoideae). Iselya 1:81-85.
Lersten, N.R. and H.T. Horner, 1976, Bacterial leaf nodule symbiosis in angiosperms with emphasis on Rubiaceae and Myrsinaceae. Bot. Rev. 42:145-214.
Malpighi, M., 1686, Opera omnia. Leiden.
Metcalfe, C.R., 1960, Anatomy of the Monocotyledons. I. Gramineae. Clarendon Press, Oxford.
Metcalfe, C.R., 1971, Anatomy of the Monocotyledons. V. Cyperaceae. Clarendon Press, Oxford.
Metcalfe, C.R. and L. Chalk, 1950, Anatomy of the Dicotyledons. 2 vols. Clarendon Press, Oxford.
Metcalfe, C.R. and L. Chalk, 1980, Anatomy of the Dicotyledons. (2nd ed). I. Clarendon Press, Oxford.
Netolitzky, F., 1932, Die Pflanzenhaare. *In:* K. Linsbauer (ed.) Handb. d. Pflanzenanatomie 4:1-253. Gebr. Bornträger, Berlin.
Orr, M.Y., 1923, The leaf glands of *Dioscorea macroura,* Harms. Notes Roy. Bot. Gard, Edinburgh, 14:57-72.
Payne, W.W., 1978, A glossary of plant hair terminology. Brittonia 30:239-255.
Prantl, K., 1891, Cruciferae. *In:* A. Engler and K. Prantl (eds.), Die natürlichen Pflanzenfamilien III, 2:145-206.
Roe, K.E., 1971, Terminology of hairs in the genus *Solanum.* Taxon 20:501-508.
Rollins, R.C. and U.C. Banerjee, 1975, Atlas of trichomes of *Lesquerella* (Cruciferae) (48 pp). Harvard Univ. Press.

Rollins, R.C. and U.C. Banerjee, 1976, Trichomes in studies of the
    Cruciferae. *In:* V.G. Vaughan, A.J. McLeod, and B.M.G.
    Jones (eds.), The Biology and Chemistry of the Cruciferae,
    pp. 145-166. Academic Press, London, New York, and San
    Francisco.
Schaede, R., 1939, Die Bakteriensymbiose von *Dioscorea macroura*.
    Jahrb. Wiss. Bot. 88:1-21.
Schnepf, E., 1969, Sekretion and Exkretion bei Pflanzen.
    Protoplasmatologia VIII, 8. Springer, Wien and New York.
Solereder, H., 1899, Systematisches Anatomie der Dikotylen.
    Enke, Stuttgart.
Solereder, H. and F.J. Mayer, 1928-1933, Systematische Anatomie
    der Monokotylen. Bornträger, Berlin.
Staudermann, W. von, 1924, Die Haare der Monokotylen. Bot.
    Archiv 8:105-184.
Theobald, W.L., J.L. Krahulik, and R.C. Rollins, 1980, Trichome
    description and classification. *In:* Metcalfe, C.R. and L.
    Chalk (eds.) Anatomy of the Dicotyledons, 2nd ed., I, 40-53,
    Clarendon Press, Oxford.
Uphoff, J.C.T., 1962, Plant hairs. *In:* W. Zimmermann and P.G.
    Ozenda (eds.), Encyclop. Plant Anatomy IV, 5:1-206. Gebr.
    Bornträger, Berlin-Nikolassee.

STRUCTURE, DEVELOPMENT AND COMPOSITION OF GLANDULAR TRICHOMES OF

CANNABIS SATIVA L.

Paul G. Mahlberg, Charles T. Hammond,[1] Jocelyn C. Turner, and John K. Hemphill

Department of Biology, Indiana University, Bloomington, Indiana 47401

ABSTRACT

The glandular secretory system of Cannabis sativa L. *is composed of bulbous, capitate-sessile, and capitate-stalked glands which are distinguishable from each other by morphogenesis and physiology. Bulbous and capitate-sessile forms occur on vegetative and floral axes whereas the highly evolved capitate-stalked form is present only on floral-related organs. In studies of cloned plants, gland initiation occurred on leaves and pistillate bracts throughout organ ontongeny. Gland density and time of appearance varied between both clones and organs, indicating that control of development is independent for each trichome type. Cannabinoid synthesis also occurred throughout organ ontongeny but with a decreasing rate in leaves as compared to an increasing rate in bracts. In individual glands, cannabinoid content decreased during maturation. Capitate-stalked glands contained higher cannabinoid levels than the sessile form although the glands maintained the profile characteristic of the clone. Analyses of glands and tissues indicated cannabinoids may occur in cells other than glands. Capitate glands develop a disc of secretory cells, and secretions accumulate in a cavity beneath a sheath derived from separation of the cuticularized outer wall surface of the disc cells. Presumed secretions, including cannabinoids, occur at the surface of plastids and appear to migrate to the cell surface adjoining the secretory cavity. Materials appear to be compartmentalized into spheres of variable size in the cavity. Cell fractionation studies are in progress to define the cannabinoid sythesizing activities within the dynamic glandular system of* Cannabis.

[1]Present address: Saint Meinrad College
St. Meinrad , Indiana 47577

1. INTRODUCTION

The epidermal glandular secretory system of *Cannabis* is an external type consisting of capitate glandular trichomes which secrete a resinous product. First descriptions of this system were made during the 19th century (Martius 1855; Unger 1866; Flückiger and Hanbury 1878; Tschirch 1889) and most extensively by Briosi and Tognini (1884). More recently, other studies utilizing both light and electron microscopy have focused on the glands for their potential utilization in *Cannabis* systematics and forensics, and for their association with cannabinoids (Bouquet 1950; Mohan Ram and Nath 1964; Shimomura 1967; Fairbairn 1972; De Pasquale 1974; Ledbetter and Krikorian 1975; Dayanandan and Kaufman 1976). Other studies have been concerned primarily with the cannabinoids, terpenophenolic compounds present only in *Cannabis*, that accumulate in the glandular trichomes (Fujita et al. 1967; Fairbairn 1972; Malingré et al. 1975; André and Vercruysse 1976). Several investigations in our laboratory have been directed specifically to studies of individual glandular trichomes (Hammond and Mahlberg 1973, 1977, 1978; Turner, Hemphill and Mahlberg 1978) as well as populations of trichomes on different organs (Turner et al. 1977, 1980). Our studies have emphasized morphogenetical and ultrastructural features of trichome development, and the analyses of cannabinoid contents of the various glandular trichomes. In addition, we have examined the interrelationships of the gland populations and the cannabinoid composition on developing plant organs. In this report we wish to bring together a number of salient features that we have found to be associated with trichome morphogenesis and cannabinoid production during the development of the glands that compose the secretory system in *Cannabis*. The objectives of our extensive studies are to determine the morphogenetic factors that control trichome development and cannabinoid synthesis, and to determine the pattern of evolution of the complex trichome system present on the different strains of *Cannabis*.

2. MATERIALS AND METHODS

Cannabis plants were specific strains (Hammond and Mahlberg 1973) or clones (Turner et al. 1977) derived and cultured as previously described. Scanning electron microscopy (SEM) was done with an ETEC Autoscan (Hammond and Mahlberg 1977; Turner et al. 1977), and SEM also was utilized for gland quantitation (Turner et al. 1977). Transmission electron microscopy was done on a Hitachi HU-11C, and samples were prepared as described previously (Hammond and Mahlberg 1978). Cannabinoid content was determined by gas-liquid chromatography (GLC) on a Hewlett-

Packard 5710A chromatograph equipped with a Hewlett-Packard 3380A integrator as described previously (Turner et al. 1977).

3. RESULTS AND DISCUSSION

3.1 *Gland description*

Three types of glandular trichomes are present on the epidermis of the outer surface of bracts from the pistillate plant of *Cannabis* as illustrated in Figure 1. The three types of glands include bulbous, capitate-sessile, and capitate-stalked (Fig. 2). Nonglandular trichomes also are present on the plant in abundance (Figs. 1,2), but will not be discussed here in detail. When mature, each of the three glandular trichome types consists of two distinct components. A secretory head comprises the top portion that is supported for each gland type by an auxiliary portion consisting of a layer of stipe cells subjacent to the head and a layer of base cells associated with the epidermis.

Bulbous glands are the smallest of the three glandular types. They are approximately 25-30 μm high with a secretory head about 20 μm in diameter (Fig. 3). The secretory head may contain one to four secretory cells in a single layer. The auxiliary portion consists of a one or two-celled base layer adjacent to the epidermis and supporting a one or two-celled stipe layer. Capitate-sessile glands have a large globose head approximately 40-60 μm in diameter (Fig. 4). Although the glands appear to be positioned directly on the epidermal surface, the secretory gland head is supported on a short axis consisting of base cell and stipe cell layers. Secretory cells within the gland head are arranged in a single discoidal layer of 8 to 13 cells bounded on the top with a membranous sheath under which accumulate the glandular secretory products. Capitate-stalked glands, also supported by an auxiliary portion, have gland heads similar in size and structure to the capitate-sessile gland heads, although they are frequently as large as 100 μm in diameter (Fig. 5). Capitate-stalked glands are distinguishable from other glands by the presence of a stalk which is derived secondarily from epidermal and subepidermal tissues rather than from immediate derivatives of the gland initial.

Bulbous and capitate-sessile glands are present on most aerial epidermal surfaces of both pistillate and staminate plants. Capitate-stalked glands are found only on flowering regions of pistillate plants, specifically on bracts and small leaves adjacent to bracts. A capitate gland also develops selectively along the clefts, but not sutures, between anther sacs of the staminate plants, and is interpreted to represent a fourth type of glandular trichome which we identify as the antherial capitate-sessile gland. This gland, although it is composed of a secretory

Key to labeling

B - bulbous gland   BC - base cell   CW - cell wall
D - dictyosome   ER - endoplasmic reticulum
F - fibrillar material   L - lipid body
M - mitochondrion   N - nonglandular trichome
Nu - nucleus   P - secretory product   Pl - plastid
S - secretory cell   SC - secretory cavity
Se - capitate-sessile gland   SP - stipe cell
St - capitate-stalked gland   V - vacuole

Figures 1-6. Morphology

    Fig. 1. Young 3 mm long pistillate bract with abundant glandular and nonglandular trichomes. x 35.
    Fig. 2. Mature bract with nonglandular trichomes and three glandular trichome types. x87.
    Fig. 3. Bulbous gland. x 540.
    Fig. 4. Capitate-sessile gland. x 810.
    Fig. 5. Capitate-stalked gland. x 810
    Fig. 6. Antherial capitate-sessile glands. x 205.

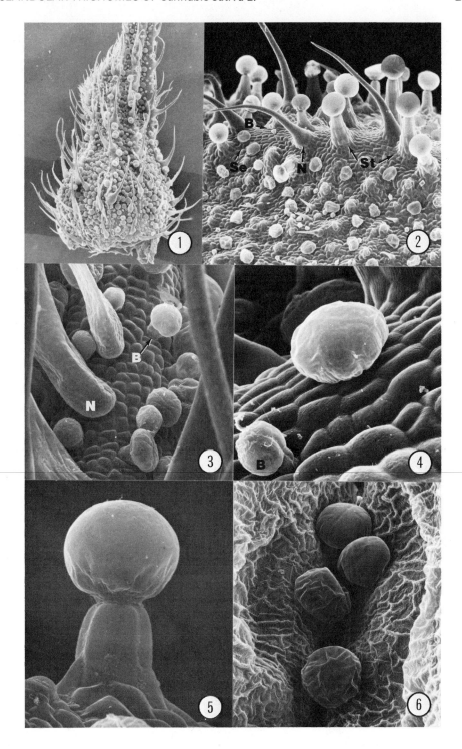

Figures 7-13. Early capitate-stalked gland development.

Fig. 7. Gland initial. x 2,250.
Fig. 8. Two-celled gland stage. x 2,045.
Fig. 9. TEM of two-celled gland stage. x 9,000
Fig. 10. Young gland following periclinal division in lower half of auxiliary portion producing basal and stipe cell layers. x 1,390.
Fig. 11. TEM of first periclinal division in gland creating upper secretory portion and lower auxiliary portion. x 4,100.
Fig. 12. TEM of periclinal division in lower half of auxiliary portion producing basal and stipe cell layers. x 3,900.
Fig. 13. Formation of flattened, four-celled gland head. x 1.905.

GLANDULAR TRICHOMES OF *Cannabis sativa* L.

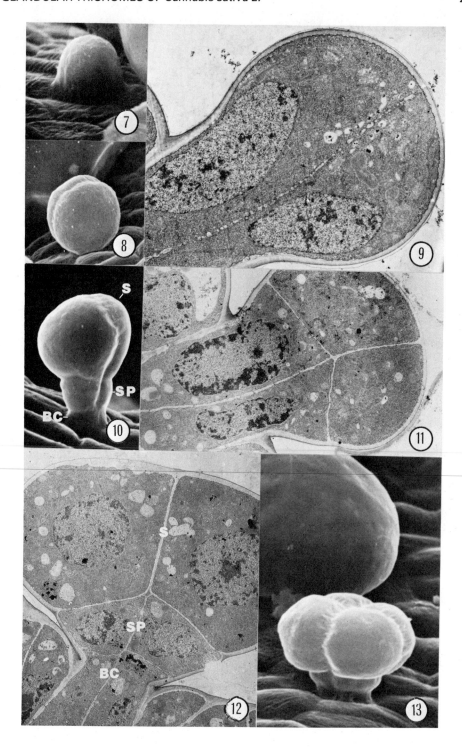

head on a short auxiliary portion, is sometimes found to be at least twice as large as any of the capitate gland types. It often develops into an oblong shape with its elongated axis parallel with the cleft of the anther. Further studies are in progress to elucidate more fully the morphogenesis and secretory content of the antherial capitate-sessile gland. Both capitate-stalked and capitate-sessile glands have been shown to contain cannabinoids. Bulbous glands also are observed to produce a secretory product, but as yet there is no direct evidence for the presence of cannabinoids in these glands. The small size of the bulbous gland makes it very difficult to collect samples from the head for chromatographic analyses.

3.2 *Development of individual glands*

At very early stages in development of individual glands, it is difficult to distinguish between the three gland types by morphological features. As the small bulbous gland matures, it can be identified by the presence of a small secretory cavity (Fig. 3). Mature bulbous glands typically consist of a two-celled base and two-celled stipe layers supporting a two to four-celled secretory head. However, we have observed base and stipe tiers each composed of only one cell. Capitate-sessile and capitate-stalked glands initially follow a similar development sequence during enlargement of the head. The head of the capitate-sessile gland appears in close proximity to the epidermis as it matures to the secretory phase (Fig. 4), whereas the head of the capitate-stalked gland becomes elevated above the dermal surface during this phase (Fig. 5).

The early stages in the initiaion of capitate-stalked glands were studied on areas of the bract where numerous such glands were known to be undergoing development. Gland initiation is first evident when a dermal initial enlarges to form a protrusion on the epidermis (Fig. 7). At a very early stage in development of this initial an anticlinal division bisects the cell in the longitudinal plane (Figs. 8,9). Large nuclei are conspicuous in each of the derived cells. A periclinal division occurs to form a lower pair of auxiliary cells and an upper or distal pair of secretory cells (Figs. 10,11). Subsequently, a second periclinal division below the first divides the auxiliary portion of the young gland into base and stipe layers (Figs. 10,12). The base tier remains a two-celled layer, while the stipe tier and the secretory tier undergo additional anticlinal divisions. The auxiliary portion is complete with the formation of a two-celled base layer and a four-celled stipe layer. The four-cells of the secretory head appear to flatten into a disc by radial enlargement and the cells divide anticlinally to produce a secretory disc of approximately 8 to 13 cells (Fig. 13). At this stage in development, the cells no longer divide but continue to increase in size (Fig. 14). During the final stages of disc enlargement, secretory activity is initiated in the disc cells and products from this activity are deposited in a space formed above the disc of secretory cells. As the head approaches its full size it becomes elevated on a multicellular stalk (Fig. 15). The stalk is formed from surrounding epidermal as well as subepidermal cells which elongate vertically to raise the gland proper--base, stipe and secretory cells--well above the surface of the bract (Figs. 15, 16). Epidermal cells in the upper portions of the stalk may undergo transverse division during their elongation. Mature capitate-stalked glands possess a dehiscence mechanism that aids in detaching the gland head from the stalk (Figs. 17-19). Abscission regions develop in both the base and stipe tiers (Fig. 17) which allows the head to separate from the stalk (Fig. 18) leaving either the base cells or stipe cells in relief (Fig. 19).

Based on morphology, development, and distribution on the plant, we consider the *Cannabis* glandular system to consist of at least four distinct and independent gland types. Although they apparently have a similar developmental sequence during their early formative phase of growth, the mature structures are quite distinguishable from each other. Thus, these glands do not represent a single morphological type arrested at various stages in development. Bulbous glands are readily identified by their small size and accumulation of secretory products in the few-celled gland head. Capitate-sessile and capitate-stalked glands differ from each other in their distribution on the plant and in the production of the stalk. Capitate-sessile glands are ubiquitous on the aerial epidermis, whereas capitate-stalked glands are observed only on bracts or adjacent leaves on the pistillate plant. Development of the stalk, although derived from cells other than the gland initial, is dependent upon some factor related to the gland proper, because stalks do not form in the absence of a gland head. The gland in some way exerts control on surrounding cells to trigger and maintain the growth of the stalk. Since capitate-stalked glands are restricted to flowering regions of pistillate plants, the physiology of this region must in some way interact with this gland type to control morphogenesis. Similarly, the selective localization of the antherial capitate-sessile gland on the reproductive structure of the staminate plant suggests that it represents a distinctive gland type in the complex trichome system of *Cannabis*. These variations in glandular types, and the presence of several types on a particular organ such as the bract, indicate that each type reflects subtle differences in the evolution of glandular trichomes in this genus.

Figures 14-19. Capitate-stalked gland development.

    Fig. 14. Stages of increasing numbers of secretory disc cells and of increasing size. x 775.
    Fig. 15. Elevation of gland on multicellular stalk. x 725.
    Fig. 16. Mature capitate-stalked glands. x 255.
    Fig. 17. Two abscission layers (arrows) in auxiliary portion of gland at juncture of gland and multicellular stalk. x 1,510.
    Fig. 18. Gland head abscission. x 300
    Fig. 19. Mature stalked gland after abscission of gland head at stipe cell layer. x 600.

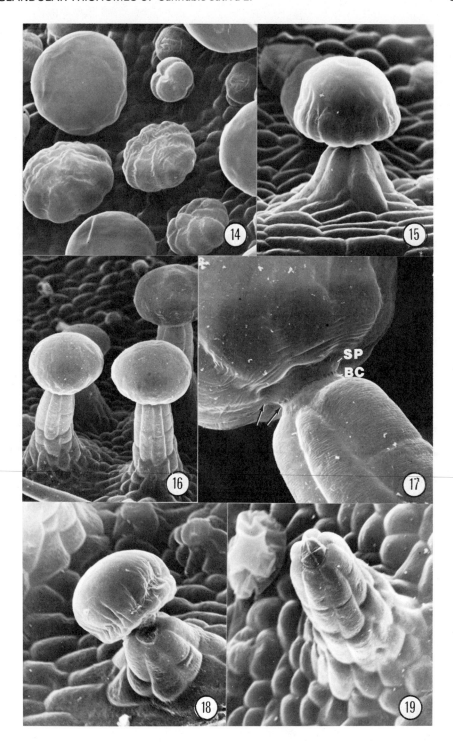

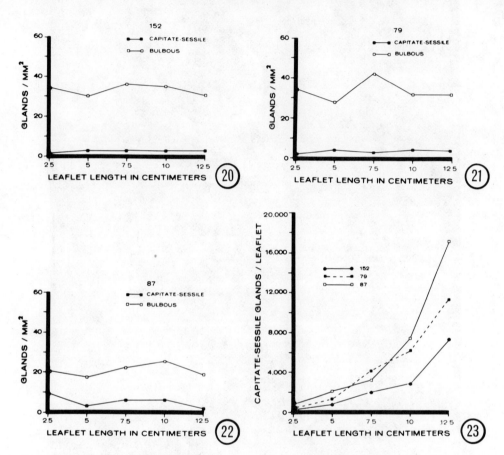

Figures 20-23. Gland populations on developing plant organs.

Fig. 20-Fig. 22. Glands per sq mm on leaves of clones 152, 79 and 87.
Fig. 23. Total capitate-sessile glands on leaves.

3.3 Gland populations on developing plant organs

Differences in patterns of gland initiation during organ development also indicated that these gland types were distinct and independent of each other. In addition, the trend of the gland population on an organ appears to depend on whether the organ is located on flowering or vegetative regions of the plant. These studies were performed on clones of *Cannabis*, each derived from a single parent plant, so as to maintain genetically uniform experimental material and to grow the plant over a multiple year period. A comparison of the gland types on the three clones, representing drug, non-drug and fiber type plants, indicated that the developing leaves and bracts of each clone possessed distinctly different populations of glands. On vegetative leaves the gland density remained relatively constant throughout leaf ontogeny (Figs. 20-22). No capitate-stalked glands were observed to develop on these leaves. Bulbous glands on all three clones were present in greater numbers than were capitate-sessile glands. Gland density remained constant on enlarging leaves and indicated that new glands were being initiated from dermal cells throughout leaf development. This is readily seen upon examining the total gland number per leaf (Fig. 23). On bracts the density of capitate-sessile and bulbous glands decreased while the density of capitate-stalked glands increased during bract ontogeny. (Figs. 24-26). Note that initiation of capitate-stalked glands was continual throughout bract ontogeny. The densities of capitate-sessile and bulbous glands decreased during bract ontogeny, although the rate of decrease for these glands was slower than the rate of increase of the surface area of the bract. Thus, new glands of each type were initiated throughout bract development. This conclusion is evident upon examination of the data illustrating the total number of glands per bract (Fig. 27).

The capitate-stalked gland was of particular interest because the time of appearance of this gland during bract development varied between clones. On the drug clone 152, capitate-stalked glands did not appear until midway through bract development (Fig. 24). This was also found to be the case for a Mexican drug strain. However, on the non-drug clone 79 (Fig. 25) and fiber clone 87 (Fig. 26), capitate-stalked glands were present on the youngest bract stages observed although in higher numbers on clone 87 than on clone 79. This temporal difference for gland initiation between clones indicates the existence of a subtle genetic mechanism which controls the initiation of this gland type. While the environment also may influence the timing of gland formation, additional studies are necessary to determine the interrelationships of genetic and environmental factors on gland initiation.

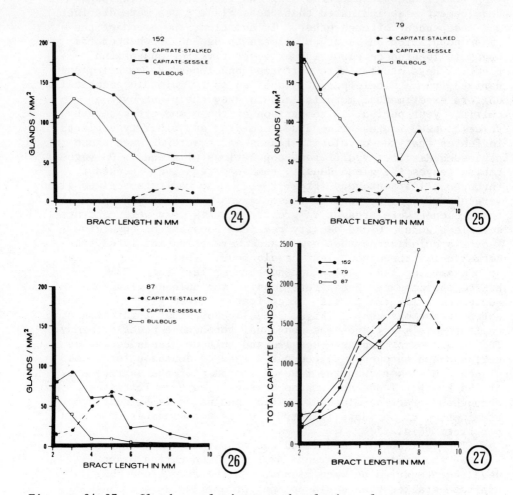

Figures 24-27. Gland populations on developing plant organs.

Fig. 24-Fig. 26. Glands per sq mm on pistillate bracts of clone 152, 79 and 87.
Fig. 27. Total capitate glands on bract.

The different patterns of the gland populations on leaves as compared to bracts may reflect the possible interaction of the physiology of the flowering region on gland type and on gland initiation. Although glands are continually initiated throughout development on both organs, the stable gland population on leaves contrasts with the progressively changing density for each gland type on the bracts. These changes in gland densities on developing bracts may be reflective of a progressive evolutionary phenomenon for the glandular system and may be related to a changing functional role of the bract during its development. Young bracts, covered with an abundance of capitate-sessile glands (Fig. 1), may protect floral organs from dessication. On old bracts, which surround developing seeds, the stratified system of bulbous, capitate-sessile and capitate-stalked glands (Fig. 2) may function to protect the maturing seed against herbivory as well as against dessication. The protective role of capitate-stalked glands on the bract is supported further by the trends in the population of nonglandular trichomes. These nonglandular trichomes are silicified, stiff hairs presumably functioning to protect the plant surface (Figs. 1,2). On clone 152, for example, no capitate-stalked glands are present on young bracts (Fig. 24) whereas the density of nonglandular trichomes is high. As the bract develops, capitate-stalked glands appear and increase in density, while the density of nonglandular trichomes progressively decreases during bract ontogeny.

3.4 *Cannabinoids in developing plant organs*

A major aspect of the glandular system of *Cannabis* is the presence of cannabinoids among the secretory products. In our initial studies (Turner et al. 1977), we started with the simple hypothesis that if specific glands were associated with cannabinoids, a correlation should exist between gland density and cannabinoid concentration of a plant organ. The results showed the system to be more complex and dynamic than originally expected. Cannabinoid content within the glands varied with gland age and, as already discussed, gland populations varied with the plant organ and developmental stage of the organ. Therefore, parallel to studying gland populations during organ development, cannabinoid content was also studied.

Cannabinoid content in leaves, based on dry weight, decreased during leaf ontogeny in the three clones studied (Figs. 28-30). The identity of the major cannabinoid component, whether Δ^9-tetrahydrocannabionol or cannabidiol, was unrelated to the cannabinoid concentration in developing leaves. All three clones showed the same decreasing trend for cannabinoids, but quantitative differences in cannabinoid concentration existed between the clones. Although the cannabinoid concentration expressed as dry weight decreased during organ development, the

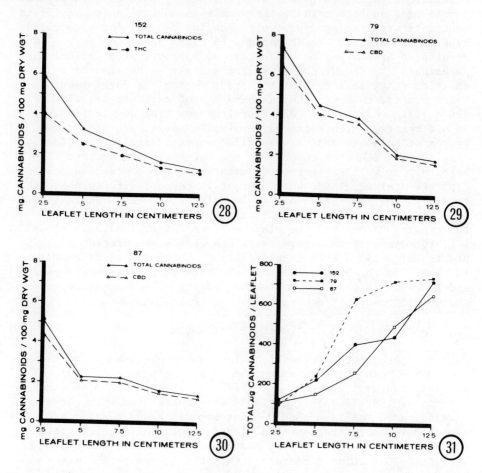

Figures 28-31. Cannabinoid content of developing plant organs.

Fig. 28-Fig. 30. Cannabinoid content, based on dry weight, in leaves of clones 152, 79 and 87.
Fig. 31. Total cannabinoids in leaves.

progressive rate of leaf expansion still resulted in an increase
in the total cannabinoid content in a leaf as it matured
(Fig. 31). All three clones showed very much the same pattern,
although clone 79 had a higher total cannabinoid content at mid-
stages in organ development when compared to the other clones.
Thus, there is continued synthesis of cannabinoids during those
stages of leaf development examined in this study.

On bracts, the trends in the cannabinoid concentrations,
based on dry weight of tissues during the ontogeny of this organ,
varied between the clones (Figs. 32-34). The drug clone, 152,
was found to have a relatively constant level of cannabinoids
throughout bract development (Fig. 32). However, non-drug clone
79 (Fig. 33) and fiber clone 87 (Fig. 34), both containing cannabidiol as the major cannabinoid, each showed an increase in
cannabinoid concentration as the bract matured. In clone 87 the
cannabinoid concentration began to increase at an earlier developmental stage of the bract than it did in clone 79. It was
apparent from these studies that cannabinoid synthesis was
occurring throughout bract development. This was readily evident
when the three clones were compared for total cannabinoids per
bract (Fig. 35).

Thus, on both leaves and bracts, it was apparent that cannabinoid synthesis occurred throughout organ development. Since
gland initiation also was found to occur throughout organ ontogeny, it was possible to determine whether a positive correlation
existed between total cannabinoids and total gland number for an
organ. Calculations based upon the total cannabinoid content distributed among the gland populations on the leaf and bract, showed
that there were differences in the cannabinoid concentration in
glands between these organs. For bracts, the estimated content
matched relatively well with actual analyses indicating that the
gland population on bracts is capable of accomodating most of
the cannabinoids present in a bract. In contrast, on leaves the
estimated cannabinoid content of an individual gland was much
higher than that found in actual analyses indicating that the
other leaf cells in addition to the glands may be producing or
accumulating cannabinoids.

3.5 *Cannabinoids in individual glands*

Looking specifically at the analyses of individual glands for
cannabinoids, several factors become apparent. An initial
finding was that cannabinoid content varied with gland age. Among
a population of capitate-stalked glands in the secretory phase,
it was possible to recognize three stages of maturation. In the
initial mature stage, the gland possessed a head with a clear
liquid content. The second stage, referred to as aged, had a
head that appeared yellow with a sticky, dense content. The final
senescent stage was represented by a gland with a red, dried head.

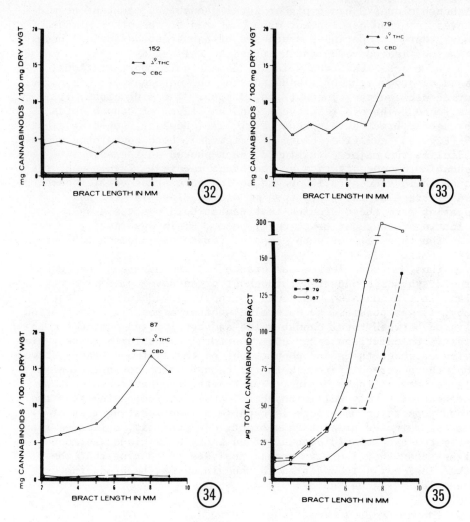

Figures 32-35. Cannabinoid content of developing plant organs.

Fig. 32-Fig. 34. Cannabinoid content, based on dry weight, in bracts of clones 152, 79 and 87.
Fig. 35. Total cannabinoids in bracts.

TABLE 1. Cannabinoid content of capitate stalked gland.

Clone	Gland age	ng Cannabinoid/gland		
		CBD	Δ^9-THC	CBN
152 (drug)	Mature	–	57	–
	Aged	–	35	21
	Senescent	–	9	1
87 (fiber)	Mature	229	–	–
	Aged	113	–	–
	Senescent	29	–	–

TABLE 2. Principal cannabinoid present in glandular trichomes of clone 152.

Gland type	Plant organ	ng Δ^9-tetrahydrocannabinol/gland		
		October	December	March
Capitate-stalked	Bract			
	Vein	55.05	60.25	43.66
	Nonvein	20.51	31.45	16.29
Capitate-sessile	Leaf			
	Vein	ncd[a]	ncd	ncd
	Nonvein	ncd	ncd	ncd (2.93[b])

[a]ncd, no cannabinoids detected
[b]100 gland sample

TABLE 3. Principal cannabinoid present in glandular trichomes of Turkish strain.

		ng cannabidiol/gland	
Gland type	Plant organ	October	December
Capitate-stalked	Bract		
	Vein	56.64	193.25
	Nonvein	102.97	145.95
Capitate-sessile	Leaf		
	Vein	8.31	ncd[a]
	Nonvein	13.42	11.72
			(11.15)[b]

[a]ncd, no cannabinoids detected
[b]100 gland sample

Cannabinoid contents decreased greatly as the gland head aged, although at each stage the cannabinoid character of the clone was maintained (Table 1). Thus, cannabinoid contents in individual glands appear to be related to the age of the gland and to the specific strain of plant.

Analyses of individual glands were expanded to include other morphological parameters for these clones as well as for several strains. Mature capitate-stalked glands from vein and nonvein areas of the bract, and capitate-sessile glands from vein and nonvein areas of the leaf, were analyzed for their cannabinoid composition. On clone 152, capitate-stalked glands from vein areas of the bract had higher levels of cannabinoids than did capitate-stalked glands from nonvein areas (Table 2). Capitate-sessile glands on leaves of 152, whether from vein or nonvein areas, contained little or no cannabinoids. A sample of 100 glands from the leaf did show the presence of cannabinoids suggesting that low levels of cannabinoids were present in capitate-sessile glands of 152. Comparable analyses of glands on clone 87 showed very similar results. On bracts and leaves of a Turkish strain, high in cannabidiol, glands from vein and nonvein areas varied but without any specific trends (Table 3). Cannabinoid levels in the Turkish strain were considerably higher than levels found in clones 152 or 87. Capitate-sessile glands generally were found to contain cannabinoids. Analyses of a Mexican strain high in Δ^9-tetrahydrocannabinol, were found to be quite similar to the analyses of the Turkish strain even though the glands differed for the principal type of cannabinoid present in the plant.

The conspicuous differences in cannabinoid concentration between the capitate-stalked and capitate-sessile glands is a character that aids to distinguish between these two gland types. However, it is possible that the differences in cannabinoid content between these gland types, as found in these studies, may be related to their location either on bracts or leaves. The larger size of the capitate-stalked gland head compared to the head of the capitate-sessile gland also may be a factor in their differing cannabinoid content. Yet other factors must be involved because the ratio for size and volume of these two gland types does not correlate with the ratio of cannabinoid levels. It is interesting to note that capitate-stalked glands have the highest cannabinoid concentration. The development of a stalk and the increased concentration of cannabinoids, therefore, indicate the evolutionary trends for the glandular system of *Cannabis*.

3.6 *Gland Ultrastructure*

While gland morphology as well as the cannabinoid content of glands have been studied in some detail by a number of investigators, little has been done to relate gland ultrastructure to the production of cannabinoids. It is well known that the formation

of secretory products in glands of *Cannabis* results in the
formation of a distended cuticular sheath area in which accumulate
the secretory products derived from the disc cells. We have
examined the ontogeny of the gland and development of the secretory cavity in an effort to determine the possible cytoplasmic
components involved in cannabinoid synthesis and the process by
which the secretory sac is formed during gland development.

In the pre-secretory stage, the head of the young capitate-
stalked gland undergoes radial enlargement to initiate formation
of the disc cells. The protoplasm of the immature disc cells
contain few endoplasmic reticulum (ER) elements, dictyosomes or
plastids (Figs. 9,11,12). The plastids in these cells contain a
poorly developed thylakoid system compared to chloroplasts in
adjacent cells. As the disc of the gland develops to the 4-celled
stage, ephemeral lipid bodies and fibrillar material become
evident (Fig. 36). Plastids at this stage frequently accumulate
an electron dense material within the stroma and at scattered
locations within the plastid envelope (Fig. 36).

As the disc cells progress from the 4-celled to the 8-13
celled stage, they develop large central vacuoles that appear to
be derived from ER. Prior to secretory activity, the large
vacuoles accumulate an electron dense material that typically
occurs along the inner surface of the vacuole membrane (Fig. 37).
Also prior to the onset of the secretion, the cytoplasm of the
disc cells becomes highly electron dense. Plastids, mitochondria,
tubular and branched ER are abundant at this stage in development,
but dictyosomes as well as dictyosome-derived secretory vesicles
are few in number (Fig. 38). A distinctive feature of the glands

Figures 36-40. Gland ultrastructure.

    Fig. 36. Four-celled disc stage, with fibrillar material and lipid bodies. x 10,300.
    Fig. 37. Secretory disc with large vacuoles that have accumulated electron dense material. x 1,200.
    Fig. 38. Secretory cells just prior to secretion. x 18,230.
    Fig. 39. Cytoplasmic connection between cells of secretory disc. x 25,750.
    Fig. 40. Formation of secretory cavity. x 3,300.

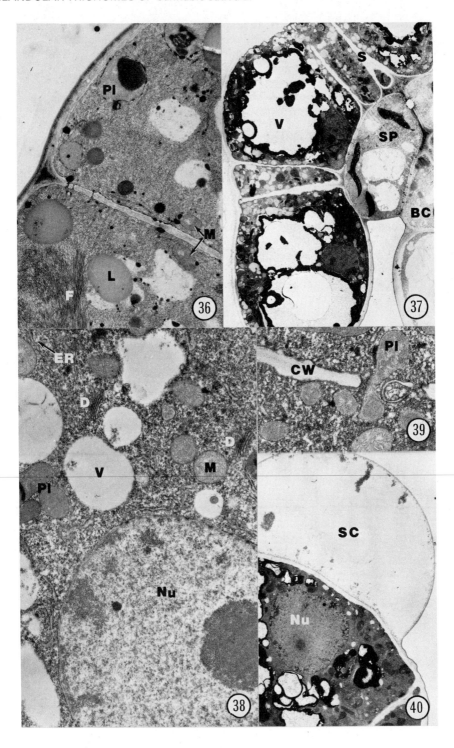

in *Cannabis* at this stage in development is the formation of a
symplast within the disc of secretory cells. Large, intercellular
cytoplasmic connections develop between secretory disc cells
(Figs. 37,39). These intercellular connections, which permit free
exchange of cytoplasm between cells of the disc, were never
observed between secretory cells and the underlying auxiliary
cells or between the auxiliary cells.

Secretory activity of the gland begins with synthesis of
secretory product and its storage in a secretory cavity above the
disc cells (Figs. 15,40). As secretory activity commences, there
are corresponding developmental changes occurring in the plastids.
While increasing greatly in number, the plastids also form complex
paracrystalline membranous inclusions (Fig. 41). At maturity, the
paracrystalline body occupies nearly all the open stroma area, the
plastids cease dividing and remain spherical in shape with a
diameter of 1.4-1.6 μm (Figs. 42, 43). Plastids containing the
paracrystalline bodies were only observed in secretory cells and
may be the potential source of the main secretory product.
Material interpreted as secretory product appears at the surface
of the plastids and then migrates through the cytoplasm to accumulate along the cell surface adjoining the secretory cavity

Figures 41-45. Gland ultrastructure.

    Fig. 41. Early secretory stage. Plastids increase in number by pinching (arrow) and begin to organize paracrystalline inclusions. x 11,000.

    Fig. 42. Late stage in secretory activity. Plastids with secretion product at their surface, stages of migration and accumulation of secretory product along cell surface adjoining secretory cavity. x 12,500.

    Fig. 43. Plastids with growing and well-developed paracrystalline bodies. x 13,300.

    Fig. 44. Formation of secretory cavity by separation in middle layer of three-layered cell wall. x 32,100.

    Fig. 45. Secretory product within the secretory cavity. x 4,900.

GLANDULAR TRICHOMES OF *Cannabis sativa L.*

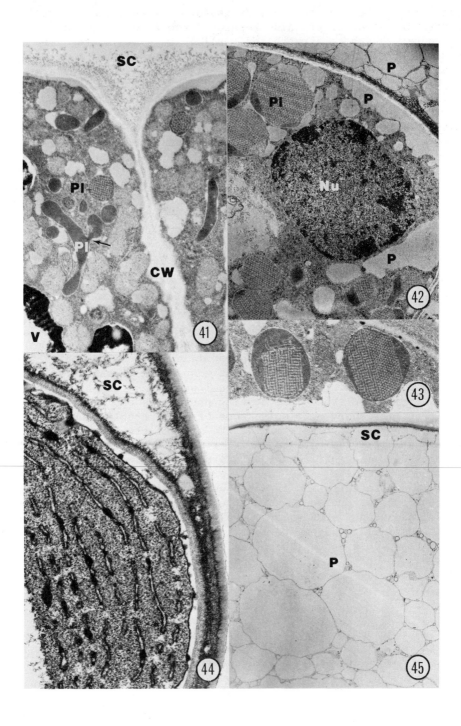

(Fig. 42). The secretory cavity is formed when the outer wall layers of the secretory disc separate from inner layers allowing the outer sheath to stretch and swell (Fig. 44). Histochemical staining of the sheath to demonstrate the presence of polysaccharide, has indicated that the outer sheath membrane appears to be comprised of both cuticle and a portion of the primary cell wall (Hammond and Mahlberg 1978). Within the secretory cavity, the secretory product can be observed to be organized into spherical bodies that seem to be delimited by a membrane-like structure (Fig. 45). No obvious features, such as pores, were observed in the wall of the disc cell that would facilitate movement of the secretory product into the secretory cavity. Since this was the case, it also was assumed that the organization of the secretory product into spherical bodies occurred in the secretory cavity rather than in the cytoplasm. Studies are in progress, using a micromanipulator, to analyse the contents of the secretory cavity not only for cannabinoids but other secretory products including proteins, lipids, or polysaccharides which may be related to the cannabinoid components.

3.7 Cell fractionation

Because the secretory product of *Cannabis* is chemically complex and contains various mono- and sesquiterpene essential oils in addition to the terpenophenolic cannabinoids, the identification of a particular product at a specific organellar site can be difficult. In addition, the lipophilic products may be extracted during typical processing procedures employed in electron microscopy. To solve some of these problems and provide another approach for determining the site of cannabinoid synthesis within the cell, we have begun fractionation studies. Our initial efforts have been directed toward preparing a fraction of viable plastids and analyzing it for the presence of cannabinoids. The preliminary results thus far obtained have indicated that no cannabinoids are present in the fraction containing viable plastids. Additional studies of the plastid fraction as well as other fractions isolated from the leaf and bract are necessary. However, it does appear that fractionation procedures may make it possible to correlate cannabinoid synthesis with a particular cytoplasmic or organellar fraction.

4. CONCLUSIONS

In conclusion, our research to this point has revealed numerous facets of the complex and dynamic glandular system of *Cannabis sativa*. Based on morphological and physiological differences, we recognize that several distinct gland types have evolved in

Cannabis. Although they may share similar ontogenetic origins, for reasons yet unknown and as a result of factors not yet know, their structure and functions differ. Also, gland populations differ as a result of their location on the plant. While initiated throughout organ development, the number of each gland type present is correlated with vegetative or flowering regions of the plant. Production of cannabinoids occurs throughout organ development but differs qualitatively and quantitatively based on a number of factors. The strain on which the glands are located dictates the qualitative aspect of the cannabinoid profile. However, gland type, gland age and location of the gland on the plant and on the organ help determine the quantitative aspects of the cannabinoid profile. Research so far has provided insight into a number of the complexities of the cannabinoid producing glandular system. Further studies should elucidate more of the controls governing gland and cannabinoid formation.

Having some information on gland and cannabinoid production and factors involved in their control, the site of cannabinoid synthesis within the gland is also of interest. Ultrastructural studies have provided an indication of plastid involvement in cannabinoid synthesis, and cell fractionalization studies are in progress to investigate this further. Once the site of synthesis is known, it should be possible to determine more fully the biosynthetic process of cannabinoid formation in the plant and perhaps be able to manipulate it.

Cannabis sativa provides a desirable system for studying plant trichomes, secondary product formation and interrelationships of the two. Distinct gland types are present and their presence is apparently related to physiological aspects of the plant. This allows an investigation of factors involved in gland initiation. In addition, genetic strains exist with apparent differences in gland populations. Secondary products produced by these glands also appear to be regulated genetically by the plant strain and environmentally by the physiological condition of the plant. *Cannabis sativa* provides a dynamic and complex secretory glandular system with variables present that allow definitive studies into the biosynthesis and control of that system.

ACKNOWLEDGEMENTS

This research was supported with grants from the National Institute on Drug Abuse (DA 00981) and the United States Department of Agriculture (53-32R6-922) to PGM. D.E.A. Registration No. PI 0043113.

5. REFERENCES

André, Cl., and A. Vercruysse, 1976, Histochemical study of the stalked glandular hairs of the female *Cannabis* plants, using Fast Blue Salt. Plant Med. 29:361-366.

Bouquet, R.J., 1950, *Cannabis*. Bul. Narcotics 2:14-30.

Briosi, G., and F. Tognini, 1894, Intorno alla anatomis della canapa *(Cannabis sativa)* Parte prima: Organi sessuali. Atti Ist. Bot. Pavia, Ser. 2-3:91-209.

Dayanandan, P., and P. Kaufman, 1976, Trichomes of *Cannabis sativa* L. (Cannabaceae). Amer. J. Bot. 63:578-591.

DePasquale, A., 1974, Ultrastructure of the *Cannabis sativa* glands. Planta Med. 25:238-248.

Fairbairn, J., 1972, The trichomes and glands of *Cannabis sativa* L. Bull. Narcotics 24:29-33.

Flückiger, F.A., and D. Hanbury, 1878, Histoire des drouges d'origine vegetale. V.2. Octave Doin, Editeur, Paris.

Fujita, M., S. Hiroko, E. Kuriyama, M. Shigehiro, and M. Akasu, 1967, Studies on *Cannabis*. II. Examination of the narcotic and its related components in hemps, crude drugs, and plant organs by gas-liquid chromatography and thin-layer chromatography. Annu. Rep. Tokyo Coll. Pharm. 17:238-242.

Hammond, C.T., and P.G. Mahlberg, 1973, Morphology of glandular hairs of *Cannabis sativa* from scanning electron microscopy. Amer. J. Bot. 60:524-528.

Hammond, C.T., and P.G. Mahlberg, 1977, Morphogenesis of capitate glandular hairs of *Cannabis sativa* L. (Cannabaceae). Amer. J. Bot. 64:1023-1031.

Hammond, C.T., and P.G. Mahlberg, 1978, Ultrastructural development of capitate glandular hairs of *Cannabis sativa* L. (Cannabaceae). Amer. J. Bot. 65:140-151.

Ledbetter, M.C. and A.D. Krikorian, 1975, Trichomes of *Cannabis sativa* as viewed with scanning electron microscope. Phytomorphology 25:166-176.

Malingré, T., H. Hendricks, S. Batterman, R. Bos, and J. Visser, 1975, The essential oil of *Cannabis sativa*. Planta Med. 28:56-61.

Martius, G., 1855, Pharmakologisch - Medicinische Studien uber den Hanf. Junge and Sohn, Erlangen.

Mohan Ram, H.Y., and R. Nath, 1964, The morphology and embryology of *Cannabis sativa* Linn. Phytomorphology 14:414-429.

Shimomura, H., M. Shigehiro, E. Kuriyama, and M. Fujita, 1967, Studies on *Cannabis*. I. Microscopial characters of their internal morphology and spodogram. Annu. Rept. Tokyo Coll. Pharm. 17:232-237.

Turner, J.C., J.K. Hemphill, and P.G. Mahlberg, 1977, Gland distribution and cannabinoid content in clones of *Cannabis sativa* L. Amer. J. Bot. 64:687-693.

Turner, J.C., J.K. Hemphill, and P.G. Mahlberg, 1978, Quantitative determination of cannabinoids in individual glandular trichomes of *Cannabis sativa* L. (Cannabaceae). Amer. J. Bot. 65:1103-1106.

Turner, J.C., J.K. Hemphill, and P.G. Mahlberg, 1980, Trichomes and cannabinoid content of developing leaves and bracts of *Cannabis sativa* L. (Cannabaceae). Amer. J. Bot: in press.

Tschirch, A., 1889, Angewandte Pflanzenanatomie. Band 1. Urban und Schwarzenberg, Wien.

Unger, F., 1866, Grundlinien der Antomie and Physiologie der Pflanzen. Bramuller, Wien.

THE SYSTEMATIC IMPLICATION OF FLAVONOIDS SECRETED BY PLANTS

Eckhard Wollenweber

Institut für Botanik der Technischen Hochschule
Schnittspahnstraße 3, 6100 Darmstadt, W. Germany

ABSTRACT

Flavonoid glycosides are very widely distributed in plants. Free flavonoid aglycones have been known for a long time to occur externally only on leaves and inflorescences of Primula *and on fronds of* Pityrogramma *species as a farinose deposit, secreted by capitate glandular trichomes. The pattern is typical for the genus* Primula; *in Gymnogramoid ferns, correlation with species and established varieties is observed as well as with chemotypes. Species-specific patterns have been found also in the bud excretions of* Populus, Aesculus, *and* Betulaceae *families. The flavonoids are secreted by a secretory epithelium or by glandular trichomes. The phenomenon of external occurrence of free flavonoid aglycones is encountered rather unexpectedly also in several other genera of herbaceous plants, mostly living in semi-arid habitats.*

1. INTRODUCTION

Flavonoid glycosides are a large group of phenolic natural products, which are widely distributed in higher plants. The occurrence of free aglycones however, is restricted and appears to be mostly correlated with the existence of secretory structures (Wollenweber & Dietz 1981). Their presence in plant exudates or in the outer layer of plant organs has been regarded as a rare phenomenon. However, it has long been known that in some exceptional cases, flavonoid aglycones may occur in considerable amounts. Piccard (1873) for example, isolated the flavones chrysin and tectochrysin from the resinous bud excretions of *Populus balsamifera* as early as 1873. Much later, in 1936, Goris & Canal reported 2',6'-dihydroxy 4'-methoxydihydrochalcone from the same material. Some years earlier Bauer and Dietrich (1933) isolated apigenin 7,4'-dimethyl ether as "the yellow pigment of birch buds". An unsubstituted flavone was found by Müeller (1915) to be a constituent of the farinose deposit on aerial parts of some species of *Primula*. Blasdale (1903) isolated from the farinose indument of the californian goldback fern *Pityrogramma triangularis* a product he called ceroptin. The structure of this compound was established in 1959 by Nilsson. The latter worker also identified the major farina constituent of two other species of *Pityrogramma* as two chalcones and two dihydrochalcones (Nilsson 1961a,b).

2. DISCUSSION

We initiated work on these flavonoid-containing plant secretions in 1969. It soon became apparent that previous workers had dealt only with the major products present in these secretions which could be readily isolated and identified. Our thin-layer chromatographic studies revealed the existence of a great variety of flavonoids in these exudates, among which were some rare and novel compounds. Most of these newly detected compounds have since been isolated and identified by spectroscopic methods.

From the resinous bud excretions of poplars more than 30 different flavonoids have been characterized and used for chemotaxonomic comparisons (Wollenweber 1975a). The results showed that together with the lipophilic material belonging to the phenylpropanoid class (e.g. cinnamoyl-cinnamate, which causes the scent in *P. balsamifera*), most species of *Populus* secrete flavonoid aglycones. The compounds chrysin, galangin and pinocembrin are encountered in many species. Their 7-O-methyl ethers as well as galangin 3-O-methyl ether and pinobanksin acetate are less abundant. Among the rarer constituents are some methyl derivatives of the commonly occurring compounds apigenin, kaempferol and quercetin, two chalcones, and a flavanone with a hydroxyl group at C-2, the first natural representative of this type. It was shown that these flavonoids occur in species-specific patterns, with small, mostly quantitative variations. Some examples are shown in Table 1.

Species-specific flavonoid patterns are also encountered in bud secretions of *Alnus, Betula* and *Ostrya* (Betulaceae) (Wollenweber 1975b), where terpenoids are the major constituents (Wollenweber 1974a). Methyl derivatives of 6-hydroxykaempferol are typical; methyl ethers of kaempferol, quercetin, apigenin and scutellarein are abundant. The highest number of constituents were found in the bud secretion of *Betula nigra* (Wollenweber 1977 and unpublished results). The abundance of the rare flavonoids from Betulaceae is presented in Table 2. The flavonoid patterns described in Wollenweber 1975b are more readily characterized when evaluated directly on polyamide TLC.

Species of *Aesculus* which produce a resinous bud secretion also exude flavonoid aglycones (Wollenweber 1974b). It has been shown that *A. hippocastanum, A. carnea, A. indica* and *A. turbinata* exhibit specific flavonoid patterns, which are composed of methyl ethers of kaempferol and quercetin (Table 3). Samples of buds of additional exudate-forming species would be highly welcome by this author for comparison purposes.

In buds of *Populus*, the lipophilic material containing the flavonoid aglycones is secreted by a glandular epithelium on the inner side of the bud scales. In Betulaceae and in *Aesculus*, however, we find instead squamiform multicellular glandular trichomes, classified as colleteres (Fig. 1). Quite another type of glandular trichome is found on farinose species of *Primula*. Leaves and inflorescences may bear numerous stalked unicellular capitate glands. The whole surface of the terminal cell is covered with the exudate material in the shape of rod or needle-like crystals. It was previously shown that this material consists mostly of unsubstituted flavone. However, we found that 5-hydroxyflavone and 2'-hydroxyflavone are also found in 18 species (Table 4)(Wollenweber 1974c). In addition, 11 species produce 5,8-dihydroxyflavone (primetin), 11 species (not necessarily the same ones) also produce 5,2'-dihydroxyflavone, and 3'4'dihydroxy-flavone is present in 14 species, but always in trace amounts. On the whole, these flavones form a pattern which is typical for the genus. Four species also exude 5,8,2'-trihydroxyflavone; the presence of 5-hydroxy, 6-methoxyflavone was confirmed only for *Primula sinensis*. It has been reported earlier that unsubstituted flavone was present also on the leaves of species of *Dionysia*, another genus of the Primulaceae (Brunswick 1922). Unfortunately, we were only able to check two herbarium fragments for this compound. In these, we indeed found

Table 1. Flavonoid patterns of some species of *Populus*. Numbers after species names denote the total number of samples analyzed. Symbols ●, O and · indicate the abundance of the compounds in these samples. "P-4" = 2,5-dihydroxy-7-methoxy-flavanone. "P-22" = 2',6' dihydroxy-4'-mellioxy dihydrochalcone.

Flavonoids \ *Populus* species	P. nigra (8)	P. deltoides (8)	P. euramericana (30)	P. euramericana cana (6)	P. koreana (3)	P. Maximoviczii (10)	P. Simonii (3)	P. trichocarpa (9)	P. trichocarpa (5)
Chrysin	●●				●	●	●		
Chrys 7-methyl ether	●	O			O O	O O	O		
Apigenin		O							
Api 4'-methyl ether		●●●	●●●	●●	●●	●●	●●	O	
Galangin	●	●●●	●●●	●●	●●	●●	●●		
Gal 3-methyl ether			●		O	O	O		
Gal 7-methyl ether									
Kaempferol 3-methyl ether		●	● · ●	●	●	●	●		
Kae 7-methyl ether						O	O		
Kae 4-methyl ether									
Quercetin 7-methyl ether	●●	●	●	●●	●	●	●		
Qu 3,7-dimethyl ether									
Qu-3,3'-dimethyl ether		●●	●●	●●	●●	●●●	●●	O	
Qu-7,3'-methyl ether		●	●●	●●	●	●	●		
Pinocembrin		●●●	●●●	●●●	●	●●●	●●		
Pin 7-methyl ether						●			
Pinobanksinacetate									
"P-4"					·	·	·	·	
"P-22"									

Table 2. Abundance of the rarer flavonoid aglycones in bud excretions of Betulaceae (n = number of species analyzed). The numbers in the table indicate the number of species in which the compounds were found.

		Alnus (n = 13)	Betula (n = 18)	Ostrya (n = 3)
Scutellarein	6,4'-dimethyl ether	6	15	–
	6,7,4'-trimethyl ether	4	1	–
Kaempferol	3,7-dimethyl ether	3	–	2
	3,4'-dimethyl ether	–	15	3
	3,7,4'-trimethyl ether	–	–	1
6-hydroxy Kaempferol	3,6-dimethyl ether	3	–	1
	6,7-dimethyl ether	–	–	1
	6,4'-dimethyl ether	5	13	–
	3,6,4'-trimethyl ether	4	17	–
	6,7,4'-trimethyl ether	–	–	1
Quercetin	3,7-dimethyl ether	4	3	–
	3,3'-dimethyl ether	1	–	1
	3,7,4'-trimethyl ether	2	3	3
	7,3',4'-trimethyl ether	3	1	1
Naringenin	7-methyl ether	–	4	–
	4'-methyl ether	–	3	–
	7,4'-dimethyl ether	–	4	–

Table 3. Flavonols found in bud excretions of 4 species of *Aesculus*.

Flavonols		*Aesculus carnea*	*Aesculus hippocasta*	*Aesculus indica*	*Aesculus turbinata*
Kaempferol					
	3'-methyl ether	+	+	+	+
	7-methyl ether	‡	‡	‡	‡
	4'-methyl ether	+			‡
	3,7-dimethyl ether				‡
	7,4'-dimethyl ether	‡	+	+	+
	3,7,4'-trimethyl ether				
Quercetin					
	3'-methyl ether	+	+	+	+
	3,3'-dimethyl ether	+			
	7,3'-dimethyl ether	+	‡	‡	‡
	7,3',4'-trimethyl ether		+	+	+
Myricetin					
	7,3',4'-trimethyl ether	+	+	+	+

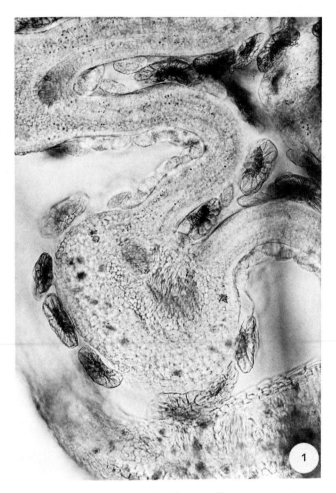

Fig. 1. Section of a bud of *Alnus glutinosa*, showing the colleters on a folded young leaf.

Table 4. The typical flavones from the farina of *Primula* and their distribution in 18 species.

Flavones substitution	Structural formulas	present in ... species (out of 18)
—		18
2'- OH		18
5,2'- OH		11
5,8,2'- OH		4
5- OH		18
5,8- OH		11
5-OH, 6- OMe		1
3',4'- OH		14

external flavonoid aglycones; unsubstituted flavone could be detected only in trace amounts.

The glandular trichomes on fronds of farinose Gymnogrammoid ferns are the same as those on farinose primroses, and the flavonoids appear in the same way, as "minute rods which are radially placed around the (glandular) cell" (Nayar 1962). This is shown in two SEM pictures (Figs. 2 and 3). Only rarely do the crystalloids form minute platelets or scales (Fig. 4). De Bary (1877) in a still unsurpassed drawing, showed what he called "pili pulverulenti" before and after treating with ethanol. The exudate sometimes forms a rather thick coating, as for example, on *Pityrogramma chrysophylla, P. lehmannii, P. triangularis* var. *pallida*. On fronds of the latter fern we found the highest amount of flavonoid exudate ever reported: 11% of the frond dry weight (Wollenweber et al. 1979). Only a few such coatings have been analyzed previously. These investigations were obviously done more or less at random, with the exception of the thorough study of Dale M. Smith and co-workers at the University of California, Santa Barbara on the *P. triangularis* complex (Star et al. 1975). Nothing was previously known on farina composition of any species of *Notholaena*, and only one species of *Cheilanthes* was investigated earlier.

We recently published a comprehensive study of some 220 samples of 14 farinose *Pityrogramma* species (Wollenweber and Dietz 1980). From this it became obvious that in this genus, as in *Primula*, we have a genus-specific flavonoid pattern. It appears that two chalcones and two dihydrochalcones with the same substitution pattern (Fig. 5) are to some extent characteristic for the genus: chalcones in goldback ferns, dihydrochalcones in silverback ferns, either alone or in mixtures. In addition, in several species we find representatives of several complex flavonoids, compounds which we designated D-1 and D-2, T-1, T-2 and T-3, X-1 and X-2 (Fig. 6)(Dietz et al. 1980) (Favre-Bonvin et al. 1980). As one can see from the structural formulas, the D and T-compounds are like "siamese twins", composed of a flavonoid moiety and a phenyl-dihydrocoumarin-moiety, linked via a common ring. The presence of these T-compounds in *P. trifoliata* together with dihydrochalcones underlines the fact that this species, which because of its different frond morphology, is treated by some authors as *Trismeria trifoliata*, should indeed be placed in *Pityrogramma* (Tryon 1962).

The one species that is different chemically from all other species of *Pityrogramma*, is *P. triangularis*, which has been the subject of detailed investigations by Dale Smith and co-workers. Within this species there exist five established varieties (var. *triangularis,* var. *pallida,* var. *semipallida,* var. *maxonii,* var. *viscosa)*, which correlate well with typical flavonoid patterns. In at least var. *triangularis* there also

Fig. 2. SEM-picture of the lower frond surface of *Pityrogramma austroamericana*. Filaments of flavonoids are found radially around the capitate glandular cell.

Fig. 4. SEM-picture of the under frond surface of *Notholaena candida var. copelandii*. The flavonoid aglycones in this case form minute plates or scales. [SEM-photos were done by Dr. W. Barthlott, Heidelberg.]

Fig. 3. SEM-picture showing the fine structure of the flavonoid filaments of *Pityrogramma austroamericana*.

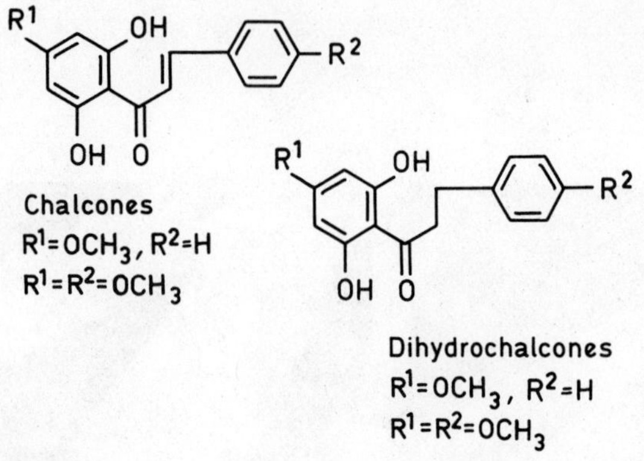

Fig. 5. Chalcones and dihydrochalcones from the farina of *Pityrogramma* spp.

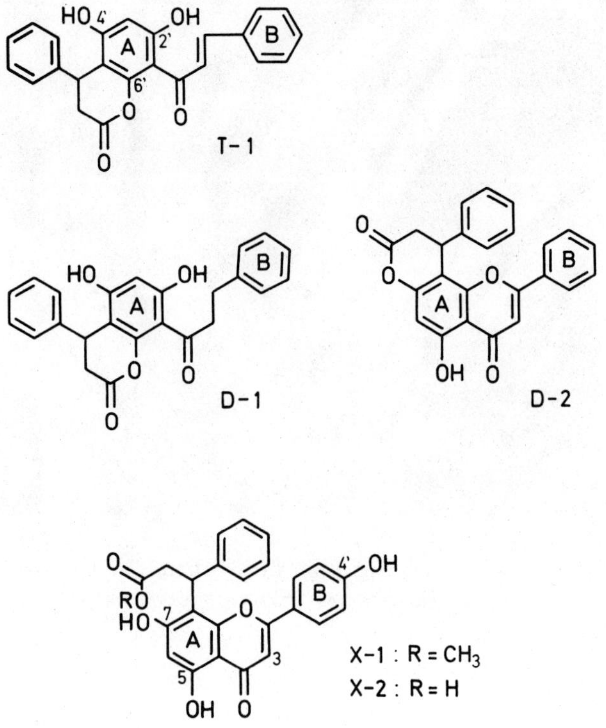

Fig. 6. New complex flavonoids isolated from the farina of *Pityrogramma* spp.

exist "chemotypes" or "chemical races", which likewise exhibit typical and constant flavonoid patterns. This has been shown by Smith (1980) and is indicated also in our article on the farina chemistry of *Pityrogramma* (Wollenweber and Dietz 1980).

Correlation of flavonoid patterns with species and with established varieties and the existence of chemotypes is observed also in *Cheilanthes* and especially in *Notholaena*. Species-specific patterns in *Notholaena* are found, for example, in *N. Grayi* (apigenin, its 7- and 4'-methyl ethers), in *N. greggii* (apigenin, its 7-, 4'- and 7,4'-methyl ethers and luteolin), in *N. sulphurea* (galangin 7 and kaempferol 7-methyl ethers, kaempferol 7,4'-dimethyl ether, 2',6'-dihydroxy-4'-methoxy-chalcone and dihydrochalcone). Examples for variety-specific patterns and chemotypes will be discussed elsewhere (Wollenweber 1981).

The occurrence of flavonoid aglycones in waxy epicuticular layers has been known for a long time in *Eucalyptus*, where some C-methylated compounds have been described by Lamberton (1964) and others. I recently discovered by chance that the thin waxy layer on the leaflets of an *Escallonia* hybrid (Saxifragaceae) cultivated in our Botanical Garden appeared to contain traces of free flavonoids mixed with ursolic acid. Since it is known that many species in this genus bear sessile glandular trichomes on their leaves and twigs, we initiated an investigation of additional species. Seven of the nine species analyzed proved to exude flavonoid aglycones (Table 5). Since these preliminary results are based on only one specimen for each species, systematic conclusions are not yet possible. These trace flavonoids are probably not constituents of the waxy epicuticular layer, but are produced by the glandular trichomes.

We also detected external leaf flavonoids on some exotic plants like *Saltera sarcocolla* (Penneacea). Chrysoeriol (luteolin 3'-methyl ether) is present in small amounts in the leaf resin of this species. We have also found the rare flavonols quercetagetin 3,6,7-trimethyl ether and quercetin 3-methyl ether (and a further, as yet unidentified flavonol) on *Barbacenia purpurea* (Velloziaceae) (Wollenweber, unpublished). This is the only member of the Monocotyledonae known to produce external flavonoids.

In previous literature little attention was paid as to the external occurrence of lipophilic flavonoid aglycones. Well known examples besides those already presented in this account are *Didymocarpus pedicellatus* (Gesneriaceae) (Seshadri 1955; Bose and Aditychaudhury 1978), *Salvia glutinosa* (Lamiaceae) (Wollenweber 1974d), and species of *Larrea* (Zygophyllaceae) (Sakakibara et al. 1976). We recently checked over three dozen plants that had been reported to accumulate flavonoid aglycones, and confirmed, at least in most cases, our assumption

Table 5. Flavonoid aglycones found externally on leaves of species of *Escallonia* (based on only one specimen each).

	Escallonia					
	illit. var. *illit.*	*myrt.* var. *myrt.*	*pulver.*	*rubra* var. *rubra*	*rubra* var. *rubra* x *illit.*	*langleyensis* (= *rubra* x *virg.*)
Chrysin						–
Apigenin	+			+	+	++
7-methyl ether		+				
7,4'-dimethyl ether			+			++
Lut						
7,3'-dimethyl ether			+			
Galangin	+		+	+	(+)	++++
3-methyl ether	+		+	+	(+)	
7-methyl ether	(+)					
3,7-dimethyl ether	(+)					
Kaempferol	+		+			(+) +
3-methyl ether	+		+	+	(+)	
7-methyl ether	+					
3,7-dimethyl ether			++			
3,4'-dimethyl ether			++			
7,4'-dimethyl ether						
Quercetin						+
7-methyl ether	+					
3'-methyl ether						
3,3'-dimethyl ether	+			+	(+)	
7,3'-dimethyl ether	+		++	++		
3,7,4'-trimethyl ether	++				+	
Pinocembrin						+

that these compounds are deposited externally. These external depositions occur in such genera as *Artemisia, Brickellia, Eupatorium, Flourensia, Haplopappus* and *Beyeria* (for preliminary chemical results see Dietz 1980). For *Flourensia, Ericameria, Encelia, Gerea* and other genera, this phenomenon has been observed earlier by E. Rodriguez and associates (pers. comm.). Urbatsch (1980) reported on the external flavonoids of some species of *Chrysothamnus*. Among the flavonoids encountered in members of the cited genera are the more or less lipophilic methyl ethers of galangin, kaempferol, quercetin, quercetagetin, chrysin, apigenin, scutellarein and luteolin. These group of substances are also known as external flavonoids from other plant sources (bud excretions, etc.). No doubt species-specific exudate flavonoid patterns exist in these plants as well. The phenomenon of flavonoid secretion seems to be most abundant in the Asteraceae, which possesses typical glandular trichomes, but it also occurs in members of the Bignoniaceae, the Euphorbiaceae, Hydrophyllaceae, Lamiaceae, Rubiaceae and other families, scattered throughout the plant system. The high number of examples of species with flavonoid exudates in Asteraceae may give a false impression, due to the low number of plants from other families analyzed so far. Also, it must be kept in mind that the Asteraceae is a very large family. Flavonoid secretion seems further to be correlated with the production of other lipophilic natural products, mainly terpenoids. We discussed these interrelations for free aglycones in a review article now in press (Wollenweber and Dietz 1981) dealing with the occurrence and distribution of some 460 free flavonoids. Most of the plants dealt with occur in semi-arid regions. Again it must be noted that the Asteraceae are abundant in such habitats. If the correlation between this phytogeographic preference and the production of external flavonoids is confirmed, there might also exist a correlation with some ecological or physiological functions. Kelsey, Reynolds and Rodriguez (Chapter 8), discuss the possible functions of such secretions.

It may be assumed that the number of free aglycones, and especially that of external aglycones, will dramatically increase in the near future, now that attention has been drawn to this phenomenon. It appears that the occurrence of flavonoids in "epidermal exudates" (like *Eucalyptus* leaf wax) is of less importance compared with that of "glandulotrop exudates" (terms used by Reznik 1959) and the secretions by glandular trichomes.

ACKNOWLEDGEMENT

Thanks are due to Dr. W. Barthlott, Heidelberg, for kind permission to publish the SEM photographs and NSF for travel support.

2. REFERENCES

Bauer, K. H. and H. Dietrich, 1933, Der Farbstoff der Birkenknospen. Ber. 66:1053-54.

Blasdale, W. C., 1903, On ceroptene, a new organic compound. J. Amer. Chem. Soc. 25:1141-1152.

Bose, P.C. and N. Adityachaudhury, 1978, Didymocarpin, a new flavanone from *Didymocarpus Pedicellata*. Phytochemistry 17:587.

Brunswick, H., 1922, Sitzber. Akad. Wiss. Wien, math.-nat. Kl., Abt. 1, 131:221.

De Bary, A., 1877, Vergleichende Anatomie der Vegetationsorgane der Phanerogame und Farne. Leipzig.

Dietz, V. H., 1980, Externe Flavonoide als Bestandteile des Farnmehls und der Epikutikularbeläge höherer Pflanzen. Ph.D. Thesis, TH Darmstadt.

Dietz, V. H., E. Wollenweber, J. Favre-Bonvin and L. D. Gomez P., 1980, Z Naturforsch. 35c:36.

Favre-Bonvin, J., M. Jay, E. Wollenweber and V. H. Dietz, 1980, Phytochemistry 19:2043

Goris, A. and H. Canal, 1936, C. R. Acad. Sci. (Paris) 201:1435.

Kelsey, R., G. Reynolds and E. Rodriguez, The chemistry of biologically active constituents secreted and stored in plant glandular trichomes, Chapter 8.

Lamberton, J. A., 1964, The occurrence of 5-hydroxy-7,4'-dimethoxy-6-methylflavone in *Eucalyptus* waxes, Austr. J. Chem. 17:692.

Müeller, H. 1915, The occurrence of flavone as the farina of the *Primula*, J. Chem. Soc. 107:872.

Nayar, B. K., 1962, Natl. Bot. Gard. Bul. Lucknow 68:1.

Nilsson, M., 1959, The structure of Ceroptene. Acta Chem. Scand. 13:750.

Nilsson, M., 1961a, Dihydrochalcones from the fronds of *Pityrogramma chrysophylla* var. maginata, domin. Acta Chem. Scand. 15:154-58.

Nilsson, M., 1961b, Chalcones from the fronds of *Pityrogramma chrysophylla* var. heyderi. Acata Chem. Scand. 15:211.

Piccard, J., 1873, Ueber das Chrysin und dessen Haloidderivate. Ber. 6:884.

Piccard, J., 1873, Ueber einige Bestandtheile der Pappelknospen. Chem. Ber. 6:890-893.

Reznik, H., 1959, In: G. Kratzl and G. Billek (eds.), Proc. 4th int. Congr. Biochem. Vienna, Vol. II, Pergamon Press, London.

Sakikabara, M., D. DiFeo, N. Nakatani, B. Timmermann and T. J. Mabry, 1976, Flavonoid methyl ethers on the ecternal leaf surface of *Larrea tridentata* and *L. divervicata*. Phytochemistry. 15:727.

Seshadri, T. R., 1955, Experientia Suppl. II.

Smith, D. M., 1980, Flavonoid anaylsis of the *Pityrogramma triangularis* complex. Bull. Torrey Bot. Club. 107:134-145.
Star, A., D. S. Seigler, T. J. Mabry and D. M. Smith, 1975, Internal flavonoid patterns of diploids and tetraploids of two exudate chemotypes of *Pityrogramma triangularis* (kaulf.) Maxon. Biochem. Syst. Ecol. 2:109.
Tryon, R., 1962, Contr. Gray Herb. 189:52.
Urbatsch, L. E., 1980, Personal Communication.
Wollenweber, E., 1974a, Zum Vorkommen von -Amyrenon in *Alnus-Arten*. Z. Naturforsch. 29c:362.
Wollenweber, E., 1974b, Methylierte flavonol-aglyka bei *Aesculus*. Z. Pflanzenphysiol. 73:277.
Wollenweber, E., 1974c, Biochem. Physiol. Pflanzen. 166:425.
Wollenweber, E., 1974d, Flavones and flavonols in exudate of *Saliva glutinosa*. Phytochemistry, 13:753.
Wollenweber, E., 1975a, Flavonoidmuster als systematisches merkmal in der gattung *Populus*. Biochem. Syst. Ecol. 3:35-45.
Wollenweber, E., 1975b, Flavonoidmuster im knospenexkret der betulaceen. ibid, pp. 47-52.
Wollenweber, E., 1977, New flavonoids from *Betula nigra*, Phytochemistry, 16:295.
Wollenweber, E., V. H. Dietz, C. D. MacNeill, G. Schilling, 1979, C-methyl-flavanones as farina on the fronds of *Pityrogramma pallida*. Z. Pflanzenphysiol. 94:241.
Wollenweber, E., V. H. Dietz, 1980a, Flavonoid patterns in the farina of goldenback and silverback ferns. Biochem. Syst. Ecol. 8:21.
Wollenweber, E., and V.H. Dietz, 1980b, Flavonoids agycones in plants. Phytochemistry. 20:869-931.
Wollenweber, E., 1981, Bot. J. Linn. Soc., in press.

HISTOCHEMISTRY OF TRICHOMES

R.L. Peterson and Janet Vermeer[1]

Department of Botany and Genetics
University of Guelph, Guelph, Ontario
Canada N1G 2W1

ABSTRACT

Plant trichomes are generally classified as covering (nonglandular) or glandular based on whether or not they function as secretory structures. The diversity of secretory products formed by glandular trichomes has provided a wide scope for the application of histochemical procedures. Identification of a variety of substances including polysaccharides, proteins, lipids, essential oils, resins and flavonoids is possible. Although most of these techniques have been employed at the light microscope level, histochemical techniques adapted for ultrastructural studies have been used in attempts to localize the sites of synthesis and storage of secretory products. In some studies, changes in nucleic acids during trichome ontogeny have been monitored histochemically. Histochemical techniques have been used to relate wall structure to function of head cells and "barrier cells" in glandular trichomes. An assessment of the results obtained by histochemical methods of covering and glandular trichomes will be included. Nectary trichomes, salt glands, and digestive glands of carnivorous plants will be excluded from the discussion. Three systems will be used to illustrate the information available through histochemical procedures: glandular trichomes located on the inflorescence of Chrysanthemum morifolium *cv Dramatic, trichomes on leaves and stems of* Pelargonium graveolens, *and paraphyses found in sori of the fern* Polypodium vulgare.

[1]Present address: Department of Biology
Carleton University, Ottawa
Ontario, Canada K1S 5B6

1. INTRODUCTION

A number or review articles have included various aspects of plant trichome structure and development (e.g. Uphof 1962; Lüttge 1971; Schnepf 1974; Juniper et al. 1977; Fahn 1979), but none have dealt specifically with histochemisty. Trichomes provide favorable and challenging objects for histochemical studies because of their diversity in form, and in the case of glandular trichomes, the variation in compounds synthesized and either secreted or stored. Trichomes often have modified cell walls and they have therefore been examined with the aid of histochemical procedures to determine the nature of wall components. A variety of histochemical procedures adapted for both the light microscope and the transmission electron microscope has been used in studies of trichomes and a considerable amount of information has been accrued using these procedures. Some of the more recent information is included in this discussion of the histochemistry of trichomes.

2. OBSERVATIONS AND DISCUSSION

2.1. *Nucleus*

Polyploidy has been considered as a possible controlling mechanism of cellular differentiation (Wardlaw 1970) but in most cases it is difficult to determine if the attainment of a polyploid state is the cause or consequence of the differentiation events in particular cells (cf Barlow 1975). Since trichomes are remarkable examples of cellular specialization, polyploidization of constituent nuclei has been studied in many species. D'Amato (1952) included trichomes among a number of cell types which undergo endopolyploidy and in a number of studies, polyploid nuclei have been recognized by comparative measurements of nuclei from various cells. Examination of Feulgen stained smears of leaf pieces of *Lesquerella* and *Physaria* species at various stages of ontogeny showed, based on increase in size of nuclei, that trichome nuclei probably become polyploid (Jakowska 1949). Trichome nuclei from numerous angiosperm species were examined for polyploid after acetocarmine staining (Tschermak-Woess and Hasitschka 1953, 1954). Although the majority of species had trichomes with diploid nuclei, a significant number had trichomes with polyploid nuclei. The most striking example of polyploid nuclei was found in the basal cells of anther trichomes of *Bryonia dioica*. The basal cell of bicellular anther trichomes of *Cucumia sativus* (Turala 1960) and *Echinocystic lobata* (Turala 1965) also has a polyploid nucleus while capitate hairs on petals of the latter species have a polyploid apical cell and diploid subtending cells (Turala 1965). Head cells of the secretory trichomes of *Achillea millefolium* (Weber and Deufel 1951) and secretory trichomes of

Solanum nigrum (Landre 1976) also have polyploid nuclei. Barlow (1975) has re-examined the anther hairs of *Bryonia dioica* and has determined, by microspectrophotometry of Feulgen stained nuclei, that the enlarged basal cell becomes as much as 256 C, whereas the distal cells are never more than 4 C. The basal cell nucleus contains polytene chromosomes which are approximately ten times the size of diploid chromosomes; homologous polytene chromosomes can apparently pair. Although the function of these anther hairs has not been determined, the large basal cell, because of its accessibility may be useful for studies relating nuclear size to cell growth (Barlow and Sargent 1975).

Sthal (1953) stained secretory trichomes of *Achillea* to follow changes in nuclear structure during hair development. The nucleus degenerates as secretion begins in these trichomes.

Nuclei of some trichome cells possess inclusions which may be fibrillar (Schnepf 1968) or crystalline (Unzelman and Healey 1972; Dell and McComb 1977). The function of these crystals is unknown but in two cases (Unzleman and Healey 1972; Dell and McComb 1977) their formation appears to be correlated with the onset of secretion, although the latter authors state that this correlation is doubtful for resin secreting trichomes in *Eremophila*. In *Pharbitis* trichomes, the intranuclear crystals are thought to be accumulations of secretory proteins (Unzleman and Healey 1972).

2.2 *Plastids*

Trichomes of various types contain chloroplasts which can be visualized at the light microscope level by bright field microscopy (e.g. Bentley and Wolf 1945; Wanstall 1950; Bancher and Holzl 1959 and see Figure 6), and by fluorescence microscopy (Stahl 1953; Dell and McComb 1975; Vermeer and Peterson 1979b). Stahl (1953) showed the loss of chloroplast fluorescence in degenerating trichome cells of *Achillea*. An example of chloroplasts in young paraphyses of *Polypodium virginianum* as seen under ultraviolet light, is shown in Figure 7. It is common for the component cells of a trichome to show differences in chloroplast numbers. For example, in *Eremophila* (Dell and McComb 1977), the foot cell of glandular trichomes contains abundant chloroplasts while the basal cell contains none. The supporting and head cells contain no chloroplasts but do have plastids with few or no internal membranes (Dell and McComb 1977). In *Solanum* trichomes both secretory and stalk cells contain chloroplasts (Bancher and Holzl 1959). Trichomes present on florets of *Chrysanthemum* have chloroplasts in all cells but those in the secretory cells lose their thylakoids as the cells mature (Vermeer and Peterson 1979b). Bentley and Wolf (1945) showed the presence of chlorophyll in tobacco trichomes by removing trichomes, treating them with 25% hydrochloric acid or glacial acetic acid at 90 C on a microscope slide and observing the formation of long, brown needle crystals that were soluble in ether or chloroform.

2.3 Crystals

Calcium oxalate crystals are ubiquitous in plant cells with trichomes being no exception. In *Eremophila fraseri*, Dell and McComb (1977) showed the presence of calcium oxalate crystals in head cells of secretory trichomes at the time of resin secretion by using a silver nitrate-hydrogen peroxide staining technique (Pizzolato 1964). In these cells, early stages of crystal deposition occur within a small membrane-bound vacuole. Crystals, which appear to be druses, are present in glandular cells of *Nicotiana* trichomes (Bentley and Wolf 1945), whereas protein crystals are present in some of the glandular trichomes of *Solanum tuberosum* (Bancher and Holzl 1959). Calcium, present as calcium carbonate depositied in the enlarged basal portion of cystolith trichomes of *Cannabis*, has been demonstrated by energy dispersive X-ray mapping techniques (Dayanandan and Kaufman 1976). Analysis of ashed samples also revealed the presence of calcium. The calcium was associated with distinct balls which were interpreted as the remains of cystoliths (Dayanandan and Kaufman 1976). Energy dispersive techniques have also been used to localize the crystalline deposits of calcium carbonate (calcite) and magnesium carbonate (nesquehonite) secreted by chalk glands of *Plumbago capensis* (Sakai 1974). The deposits from these glands, when viewed in the cathode luminescence mode of the SEM, appear to absorb electrons and emit light in the visible spectrum (Sakai 1974).

2.4 Phenols and tannins

Recent reviews (Levin 1973; Johnson 1975) emphasize the role trichomes may play in providing a defense mechanisms for plants. Among the modes of defense is included the possibility that simple or complex phenolics in trichomes may act as gustatory repellents.

Few histochemical studies of phenol localization in trichomes have been published. The most thorough consideration is the work by Beckman et al. (1972). Of the thirty-nine angiosperm species bearing trichomes examined, thirty-two had nitroso-positive substances in their trichomes. Granular or globular bodies appeared to be stained first by this staining reaction for catechol phenols. These authors studied the secretory trichomes of *Lycopersicon esculentum* (tomato) in more detail and found that the contents of capitate cells stained green with ferric chloride, cherry red with the nitroso reaction, yellow with ammonia fumes and blue with 2,6-dichloroquinone 4-chloroimine, showing quite conclusively that the substances are phenolic in nature. Correlated chromatography of epidermal strip extracts of tomato showed a large spot which reacted to the histochemical tests used on fresh hairs. These authors suggest that phenol-containing trichomes may act as sensory organs to trigger defense reactions following injury.

Phenolic substances have been identified in the unusual paraphyses (see Figure 3) found in the sorus of the fern, *Polypodium virginianum* (Peterson and Kott 1974) using the nitroso reagents (see also Figure 8). In female *Cannabis* plants, cannabinoids containing phenol groups have been localized in multicellular, stalked glandular trichomes and in sessile trichomes by the use of Fast blue B salt (Andre and Vercruysse 1976), a reagent often used to identify phenol groups in thin layer chromatography (Stahl 1966). Cannabinoids were not observed external to the cuticle of these hairs unless the cuticle had been damaged. Tannin deposits in both simple and stellate covering trichomes of *Grewia* were mentioned (Tiwari 1978) but no histochemical evidence was given.

2.5 *Proteins*

Protein can be deposited in trichomes in crystalline form, for example the nuclear crystals in *Pharbitis nil* (Unzelman and Healey 1972) and *Eremophila fraseri* (Dell and McComb 1977), or more commonly as a carbohydrate-protein complex. In trichomes of *Pharbitis nil* numerous storage vesicles containing protein, as demonstrated by staining with mercuric bromophenol blue and ninhydrin-Schiff's reagent, and carbohydrate, as shown by the PAS reaction, are found in the secretory cap cells (Unzelman and Healey 1974). The secretory product lying external to the glandular cells also stains positively for these substances. A mechanism involving coated vesicles has been proposed in the secretion of this proteinaceous meterial.

The mucilage produced by gynoecium papillae of *Apteria cordifolia* also stains for proteins using the ninhydrin technique, and for carbohydrates using the PAS technique (Kristen 1976), as does the mucilage produced by stipular trichomes of the tropical shrub *Psychotria bacteriophila* (Horner and Lersten 1968). In the latter species, the mucilaginous substance bathes the apical meristem, developing leaves, and stipules and is apparently used as a carbon and nitrogen source for bacteria which eventually move through stomata and initiate leaf nodules (Lersten and Horner 1967). After nodule initiation, secretory trichomes degenerate and are replaced by non-secretory multicellular trichomes (Horner and Lersten 1968).

The petiolar glands of *Mercurialis* have been shown to contain protein inclusions and protein-containing secretions by reacting fixed tissues with iron hematoxylin, ninhydrin, and bromophenol blue, and carbohydrates by PAS staining for light microscopy and by the periodic acid-methanine-silver reaction for transmission electron microscopy (Figier 1968). In an attempt to determine the sites of synthesis and transport of these protein compounds,

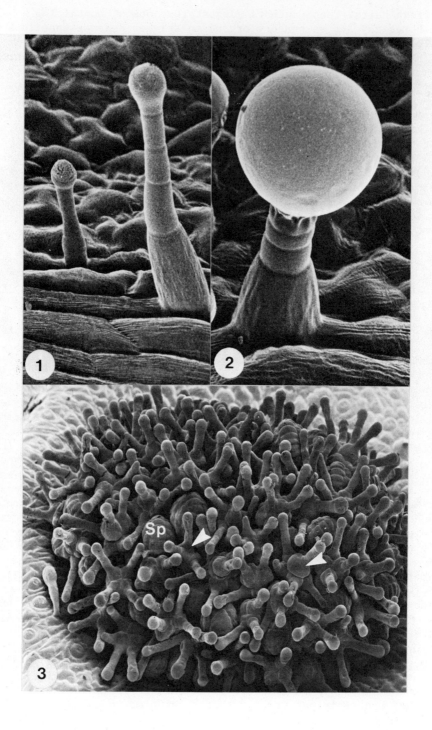

HISTOCHEMISTRY OF TRICHOMES

Figures 1-3. Scanning electron microscopy of trichomes.

Fig. 1. Glandular trichomes on leaves of rose scented geranium *(Pelargonium graveolens)*. X 70.
Fig. 2. Globular glandular trichome on leaf surface of rose scented geranium. X 120.
Fig. 3. Sorus of *Polypodium virginianum* L. showing paraphyses (arrows) and sporangia (Sp). X 110.

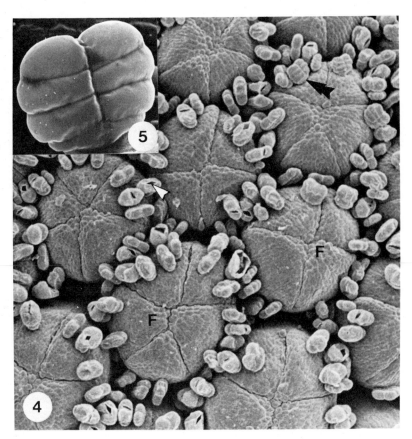

Figures 4-5. Scanning electron microscopy of *Chyrsanthemum morifolium* cv. *Dramatic* florets with glandular trichomes.

Fig. 4. Florets (F) with glandular trichomes (double arrow). An opening in the cuticle (single arrow) is evident in some trichomes. X 70.
Fig. 5. Enlargement of single glandular trichome. X 90.

Figures 6-13. Trichomes from fresh tissue.

Fig. 6. Glandular trichome of *Chrysanthemum* viewed with crossed polarizers. The outer tangential cell wall (arrows) of the upper tier of cells is isotropic. Chloroplasts are evident in some of the cells. X 270.

Fig. 7. Chloroplasts of paraphyses (P) and sporangia (Sp) scraped from a sorus of *Polypodium* fluoresence red under ultraviolet light. X 75.

Fig. 8. Large vesticulated cells (*) of *Polypodium* paraphyses showing nitroso-positive phenolics within the cells. X 260.

Fig. 9. Glandular trichomes of *Chrysanthemum* viewed with blue light. Contents of the trichome fluoresence yellow. X 90.

Fig. 10. Glandular trichomes of *Chrysanthemum* viewed under ultraviolet light. Contents of glandular trichome fluoresence silver-blue. X 80.

Fig. 11. Old glandular trichomes of *Chrysanthemum* after release of secretory product. A few crystalline deposits (arrows) remain. X 110.

Fig. 12. Top view of *Chrysanthemum* glandular trichome after staining with Sudan Black B. Lipophilic substances (arrows) are present. X 90.

Fig. 13. Globular glandular trichome of *Pelargonium* viewed with ultraviolet light. The cuticle (arrow) and substances (double arrow) within the head cell fluoresce. X 85.

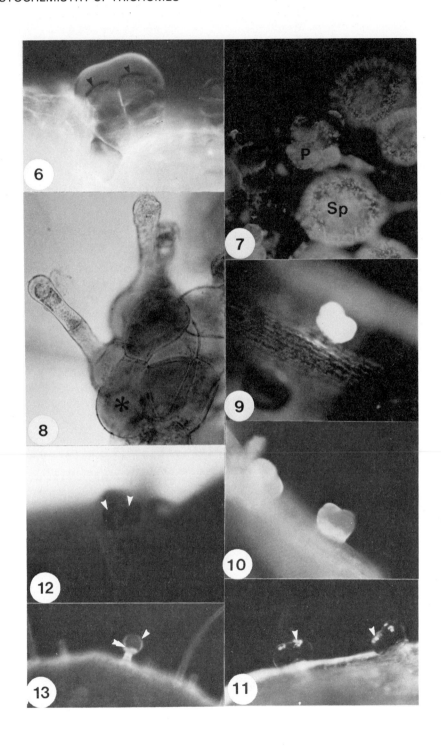

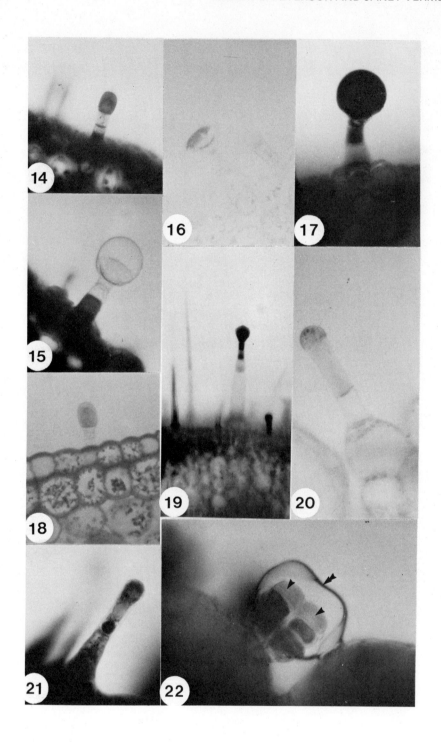

Figures 14-22. Histochemistry of trichomes. Tissue represented in Figs. 16 and 18 was embedded in glycol methacrylate prior to staining. All other trichomes are from fresh tissue.

Fig. 14. Glandular trichome of *Pelargonium* stained with PAS. The terminal cell is PAS-positive. X 160.

Fig. 15. Globular glandular trichomes of *Pelargonium* stained with PAS. The basal cell is PAS-positive but the glandular head is not. X 260.

Fig. 16. Globular glandular trichome of *Pelargonium* stained with Nile blue. The glandular head cell stains blue indicating the presence of lipophilic substances. X 260.

Fig. 17. Globular glandular trichome of *Pelargonium* treated with osmium tetroxide. Contents of the globular head cell are osmiophilic suggesting lipophilic substances. X 270.

Fig. 18. Short glandular trichome of *Pelargonium* stained with Nile blue. The head cell contains lipophilic substances. X 260.

Fig. 19. Short and tall glandular trichomes of *Pelargonium* showing osmiophilic contents of various cells in each trichome. X 100.

Fig. 20. Glandular cells of *Pelargonium* paraphyses showing pink and blue staining reactions with Nile blue. Acidic and neutral lipids are therefore present. X 270.

Fig. 21. Glandular cell similar to that shown in Fig. 20 after treatment with osmium tetroxide. The contents that reacted with Nile blkue in the previous figure are osmiophilic. X 150.

Fig. 22. Glandular trichome of *Chrysanthemum* stained with Sudan black B. The cuticle (double arrow) and particles in the glandular cells (single arrows) are stained. X 270.

labelling with ^{3}H-glycine for various times followed by autoradiography at the TEM level was employed (Figier 1969). The results showed that activity first appeared in the mitochondria, plastids, nucleus and in discrete storage vesicles in the cytoplasm. Some of the protein of the ergastoplasm appears to be passed from rough endoplasmic reticulum to Golgi bodies, smooth ER, and then to membranous whorls before being discharged as protein-containing secretory substances. Figier (1969) hypothesized that the Golgi might be involved in binding the protein to polysaccharides prior to the release of the secretory product. This would support the histochemical evidence for the presence of carbohydrate-protein materials in the secretory product (Figier 1968). Although the sequence of events envisioned by Figier (1969) may be real, one must always use caution when interpreting dynamic events from static electron micrographs (O'Brien 1972).

2.6 *Carbohydrates, polysaccharides*

Trichomes secreting a carbohydrate-based product are grouped with the hydrophilic type of gland according to Schnepf (1974). This type of trichome is found in a number of plants from mosses (Wanstall 1950; Hebant and Bonnot 1974) to higher plants (Schnepf 1974). In mosses, the group *Polytrichiales* has well-developed mucilage hairs which secrete masses of PAS-positive material over the shoot apex and developing leaves (Bonnot and Hébant 1970; Hébant and Bonnot 1974). These mucliage hairs contain numerous dictyosomes whose vesicles stain positively, at the ultrastructural level, with the periodic acid-thiocarbohydrazide-silver proteinate test for polysaccharides (Bonnot and Hébant 1970; Hébant and Bonnot 1974). In angiosperms, there are a number of reports of carbohydrate-based substances being secreted by trichomes (e.g. Figier 1968; Schnepf 1968; Rougier 1966,1972) and in most cases dictyosomes have been implicated by histochemical means to be involved in the process (Schnepf 1974). Rougier (1972) used a variety of histochemical techniques including PAS, alcian blue and colloidal iron to identify polysaccharides in squamulae (glandular scales) of *Elodea* at the light microscope level. She also showed, by the use of ruthenium red and periodic acid-thiocarbohydrazide-silver techniques for the identification of polysaccharides at the ultrastructural level, that these materials are secreted by the Golgi apparatus and pass into the cell wall by a mechanism of reversed pinocytosis. In this case, a part of the secreted material is probably pectinaceous as determined by staining with alcian blue after pectinase treatement.

The kinetics of slime production by glands of *Mimulus tilingii* has been determined (Schnepf and Busch 1976). It was concluded that one Golgi vesicle exists for about 30 minutes, Golgi vesicles occupy an average about 18.2% of the mean volume of each secretory cell, and to produce the observed volume of 34.4 µm^3 slime per minute, approximately 530 vesicles would fuse with the plasmalemma every minute. Unzelman and Healey (1974) suggest that coated vesicles may be involved in the secretion of protein-carbohydrate complexes in glandular trichomes of *Pharbitis nil*.

Figure 14 shows PAS-positive material in the small glandular trichomes (see Figure 1) of the rose scented geranium, *Pelargonium graveolens*. The large globular secretory trichomes (Figure 3) contain PAS-positive material in the stalk but not the head cell (Figure 15) although this may be due to a blockage of transport of the reactive chemicals.

2.7 *Lipophilic substances*

Lipophilic substances secreted by trichomes consist of a variety of compounds including terpenes, essential oils, resins, waxes, fats and flavonoid aglycones (Schnepf 1974). There are often mixtures of these compounds in the secretory product of trichomes (Dell and McComb 1978) making it difficult to identify the precise chemical nature of the products by histochemical procedures. More frequently, broad classes of substances can be determined histochemically. For example, in the glandular trichomes found on florets of *Chrysanthemum morifolium* cv. *Dramatic* inflorescences (Vermeer and Peterson 1979a, see Figures 4 and 5) the secretory product fluorescences bright yellow under blue light (Figure 9), silvery blue under ultraviolet light (Figure 10), and stains intensely with Sudan black B (Figure 12) and Nile blue. The product does not stain with PAS. Very old trichomes which have probably finished secreting, have a few residual crystalline deposits left in the sub-cuticular space (Figure 11). In *Pelargonium graveolens,* the globular petiolar glandular trichomes show autofluorescence under ultraviolet light (Figure 13), stain with Nile blue (Figure 16) and react strongly with osmium tetroxide (Figure 17), indicating the lipophilic nature of the secretory product. The short secretory trichomes are also positively stained with Nile blue (Figure 18) and are osmiophilic (Figure 19). Since these trichomes are PAS-positive as well (see Figure 14), a mixture of lipophilic and hydrophilic substances may be present.

The glandular cells of the paraphyses found within the sorus of the fern, *Polypodium virginianum* (see Figure 3) stain a variety of colors with Nile blue indicating both acidic and neutral lipids (Figure 20). These cells also react with osmium tetroxide in areas which correspond to the areas stained with Nile blue (Figure 21). Without the correlated information from thin layer or gas chromatography, however, nothing can be said about the chemical nature of these lipophilic substances.

2.8 *Essential oils*

Essential oils are volatile oils which are immisicible with water but soluble in organic solvents (Dell and McComb 1978). Volatile components of glandular trichomes may act as repellants to insects (Levin 1973). These compounds have been studied by thin layer or gas chromatography and a number of constituent chemicals have been identified (see Amelunxen et al. 1969; Wollenweber and Schnepf 1970; Heinrich 1973). In many trichome systems studied, the oils apparently accumulate below the cuticle in a sub-cuticular space and can be stained by various histochemical procedures. Oil-secreting trichomes of *Monarda fistulosa* show Sudan III-positive and electron dense substances below the cuticle (Heinrich 1977); substances with the same electron density were found in the mitochondria. Lassanyi et al. (1978) showed the accumulation of volatile oils within the sub-cuticular space in trichomes of *Chamomile* with the De Chatellier reagent, which contains among other things, Sudan III. Other staining techniques and cytophotometric methods were used also to show the variation in prochamazulene content in secretory cells. Based on orange staining with Rhodamin B, gold-yellow autofluorescence under utraviolet light and lack of staining with Sudan III (although why there should be no staining with this dye is not known), the secretory products of *Solanum* trichomes are considered to be an oil which perhaps contains some terpenes (Bancher and Holzl 1959). Proazulene has been identified in the subcuticular space of *Achillea* trichomes by treating tissue with orthophosphoric acid or methyl alcohol sodium methyl solution and observing the color produced (Stahl 1953). Trichomes containing proazulene are either neutral or weakly basic as shown by staining with neutral red (Stahl 1953).

The presence of oil droplets in all cells of the glandular trichomes of *Valeriana* as visualized by toluidine blue O---iodine staining, was taken as evidence that all cells are secretory (Szentpetery et al. 1969). However, since the ordinary epidermal cells also contained oil droplets, this is not very good evidence. These authors also claim that these trichomes begin to degenerate from the stalk, with the head cells being the last to degenerate; this is contrary to all other systems described.

There is some evidence that the nature of the essential oil produced in glandular trichomes can become modified with plant age. In *Ocimum gratissimum* seedlings had methyl eugenol as the major component of essential oil, whereas older plants had little methyl eugenol but more terpenes, particularly thymol (Dro and Hefendehl 1973).

2.9 Fats

Lipid deposits that are discrete solid bodies have been identified in the glandular trichomes of *Chrysanthemum morifolium* cv. *Dramatic* either by staining fresh or fixed material embedded in Spurr's resin with Sudan black B (Vermeer and Peterson 1979b). Ultrastructural evidence suggests that plastids may be involved in the synthesis of these bodies. Fats have also been demonstrated in the long and short secretory trichomes of tobacco by staining with Sudan III/IV (Akers et al. 1978), and in both erect capitate and procumbent glandular trichomes in *Medicago* by Sudan IV staining (Kreitner and Sorensen 1979).

2.10 Resins

An excellent discussion of the chemistry of resins and the structure of resin-secreting glands is found in Dell and McComb (1978). These authors point out the difficulty in locating the sites of resin synthesis with histochemical techniques, and in distinguishing between the many different components of resin. Flavonoid aglycones, water-insoluble flavonoid components of resin, show primary fluorescence under ultraviolet light, a characteristic used to identify them in the lipophilic secretions from buds of *Aesculus hippocastanum* L. (Charriere-Ladreix 1975). Resins in glandular cells of trichomes of *Newcastelia viscida* fluoresence yellow under ultraviolet light (Dell and McComb 1975). The lipophilous secretion from the calyx glands of *Plumbago* stains with Sudan IV and is red with Nile blue, indicating to Rachmilevitz and Joel (1976) that the substance is a resin. The staining reactions, however, could indicate other lipophilic materials.

Flavone secreting glands of the fern *Pityrogramma* show osmiophilic deposits in the ER and in the cisternae at the forming face of Golgi bodies (Schnepf and Klasova 1972). These flavonoids pass through the cell wall and cuticle and crystallize on the surface of secretory glands.

2.11 Enzymes

Digestive glands of carnivorous plants have received the most attention in terms of enzyme activity (see Juniper et al. 1977). Glandular trichomes of *Achillea* show considerable dehydrogenase activity as shown by formation of formazan compounds after treatment with 2,3,5-triphenyltetrazolium chloride (Stahl 1957). Mucilage secreting hairs on the gametophytes of mosses of the order *Polytrichales* are particularly active cells as shown by their ultrastructure (Bonnot and Hébant 1970). Histochemical studies have shown that these hairs show a considerable amount of reaction product when tested for the respiratory enzymes, succinic dehydrogenase and cytochrome oxidase and a number of phosphates (Hebant and Bonnot 1974). At the ultrastructural level, inositol diphosphatase activity was found in the ER, dictyosomes, small vacuoles, walls and extruded mucilage. Although doubting the specificity of the Gomori method for acid phosphatase localization at the ultrastructural level, these authors did find acitivity in the dictyosomes. Rougier (1972) also found acid phosphatase activity associated with Golgi bodies and vacuoles of mature and senescing cells in squamules of *Elodea,* and suggested a lysosomal function. Stahl (1957) reported that nuclei in glandular trichomes of *Achillea* contained Gomori-positive substances which were presumably phosphatases.

2.12 *Cell Wall*

Uphof (1962) has discussed the nature of the cell walls of trichomes in considerable detail. The basic structural component of trichome cell walls is cellulose which is generally organized in a primary cell wall (Bentley and Wolf 1945; Uphof 1962) but cell walls of many trichomes become modified either by the elaboration of additional wall layers or by the deposition of various encrusting substances. In a study of cleared leaves of fifty-two families of dicotyledonous angiosperms, unusual wall thickenings in trichomes of three families, Rosaceae, Betulaceae and Rubiaceae were described (Lersten and Curtis 1977). These helical thickenings, which stained intensely with chlorozol black E, a general stain for cellulose, were found as an outer wall layer. The authors speculate that these bands may represent the original primary cell wall which is split and stretched during cell elongation. Pits develop in the cell walls of *Grewia* trichomes, indicating that a secondary cell wall is deposited (Tiwari 1978). Although this author claims that the cell walls are lignified, no histochemical evidence was presented.

The most common encrusting substance of trichome cell walls is cutin. In many glandular trichomes, cutinization of the lateral and/or transverse cell walls of basal or stalk cells occurs (Amelunxen 1965; Schnepf 1968; Dell and McComb 1975, 1977; Lyshede 1976) presumably preventing apoplastic flow through these walls and perhaps isolating the head cells from the epidermis (Dell and McComb 1975). Also, these modified cell walls may prevent the backflow of discharged materials or reduce transpirational water loss through trichomes (Schnepf 1974). Although little is known concerning the mechanism by which cutin is deposited within trichome cell walls, Olesen (1979) has recently shown that in salt glands of *Frankenia,* thin lamellae released from the plasmalemma, as observed in glutaraldehyde-tannic acid fixed tissue, are involved in cutin deposition.

Dell and McComb (1977) showed that in *Eremophila*, the lateral cell walls of stalk cells and the outer cell wall of head cells of resin-secreting trichomes stained poorly with cationic dyes and reacted negatively with the silver hexamine reaction for polysaccharides; they suggest that the polysaccharide reaction sites are covered by cutin. Stahl (1952) observed that the cell wall of secretory cells in trichomes of *Achillea* gave only weak staining reactions for cellulose. He also showed with ruthenium red staining that a distinct pectin layer was present. Stalk cell walls of the mucilage hairs of the moss *Polytrichium* are resistant to sulphuric acid indicating that they may be cutinized (Wanstall 1950). In *Chrysanthemum* glandular trichomes, the lateral walls of stalk cells are PAS-positive but also stain positively with the lipid stains Nile blue and Sudan black B (Vermeer and Peterson 1979b). Cutin is most likely present in stalk cell walls but the positive reaction of the outer tangential wall to lipid stains is most likely due to the presence of lipophilic secretory product. The outer tangential cell wall is isotropic whereas all the other trichome cell walls are anisotropic (see Figure 6). It has been shown by electron microscopy that the outer tangential cell wall develops wall ingrowths which may have less organized cellulose microfibrils than the primary cell wall (Vermeer and Peterson 1979b). Also, the primary cell wall becomes modified by the deposition of electron-dense materials which may account for the loss of birefringence.

Elemental analysis using X-ray spectrometry has shown that all nonglandular hairs of *Cannabis sativa* L. have highly silicified cell walls (Dayanandan and Kaufman 1976). These authors have found that trichomes in *Humulus lupulus* and *Lantana* can be distinguished from those found in *Cannabis* by, among other features, a difference in X-ray patterns. *Humulus* trichomes are poorly silicified whereas epidermal cells in a ring around the base of each hair are highly silicified. In *Lantana*, a ring of basal cells is silicified while the trichome proper is calcified. It is concluded that, by the use of SEM combined with X-ray analysis, the identification of marihuana fragments even in ash remains should be possible.

2.13 *Cuticle*

Most trichomes have a cuticular covering which may vary considerably in thickness and structure depending on the position of the cell in the hair (Uphof 1962). Mucilage secreting trichomes in *Psychotria* apparently lack a cuticle (Horner and Lersten 1968). In secretory trichomes, the cuticle may remain intact until secretory product accumulates in a sub-cuticular space (e.g. Dell and McComb 1975) or pores may be present which allow secretory products to be released (Wollenweber and Schnepf 1970; Vermeer and

Peterson 1979a,b). Pores in some cases may be artifacts due to fixation techniques. There has been some question as to how the sub-cuticular space forms in glandular trichomes. Amelunxen (1964,1965) claims that in the secretory trichomes of *Mentha piperita*, the space is formed by a splitting of the cell wall between the pectic and cuticular layers. In capitate glands of *Cannabis*, the "sheath" surrounding the head region is composed of an outer layer, presumably cuticule, and an inner layer which stains positively for polysaccharides using the periodic acid-silver methanamine method (Hammond and Mahlberg 1978). This evidence is interpreted as indicating that the cell wall is cleaved in some way during the formation of the subcuticular space.

The presence of a cuticle surrounding trichomes can be demonstrated by a number of histochemical techniques. A thorough study of the cuticle of *Achillea millefolium* trichomes (Stahl 1953) showed that it did not react with Congo red, methylene blue, ruthenium red or phloroglucinol-HCI. However, it was yellow with chlorizinc iodine and hydriodic acid, weak orange with Sudan III, and weak red with scarlet red. The cuticle was insoluble in cupric oxide ammonia and dilute hot mineral acid, dissolved except for a thin film when treated with concentrated sulfuric acid and 50% chromic acid and completely dissolved in concentrated potash. These results apparently showed that the cuticle consists of two components. In *Nicotiana*, glandular trichomes, the cuticle stains with Sudan IV (Bentley and Wolf 1945) and in *Chrysanthemum*, the cuticle stains with Sudan black B (see Figure 22).

Trichomes of desert plants have been implicated in water absorption (Uphof 1962) and Lyshede (1976) has considered this possibility in a recent study of hairs of *Spartocytisus filipes*. The trichomes of this plant are T-shaped consisting of three cells, an apical cell, stalk cell and basal cell. The apical cell has a thin cuticle with small cuticular warts while the stalk cell is surrounded by a thick cutinized cell wall. Rhodamine B, a lipophilous fluorescent tracer, was used to study transport into the cells of the hair. It was concluded, based on the appearance of the fluorochrome in the cutinized cell walls of trichomes and in the xylem adjacent to trichomes, that absorption takes place through the hairs. These results are used by the author to support the concept that trichomes are able to absorb water. This extrapolation must, however, be questioned since the fluorochrome used is a lipophilous substance which would be expected to penetrate the cuticle and cutinized cell walls. These layers would be expected to be a barrier to the apoplastic movement of water.

3. CONCLUSIONS

Although a considerable amount of information has been published on the histochemistry of trichomes, some caution must be used in interpreting the results. It is doubtful whether many of the histochemical techniques used are specific enough for the determination of precise chemical compounds present in the secretory products of glandular trichomes. These products are bound to be rather complex in nature. However, most procedures can provide useful information concerning classes of compounds present. Most of the techniques used are standard botanical histochemical procedures for compounds such as polysaccharides, proteins, and lipids, but even with these the chemical basis of the staining reaction is often poorly understood, and the specificity of the method may be questioned. For example, the use of osmium tetroxide to indicate lipids may be misleading since some phenolic compounds show similar reactions (Nielson and Griffith 1978). Likewise, the well-known periodic acid-Schiff's reaction for polysaccharides with 1,2 vicinal groups, has recently been shown to give positive results with some phenolic inclusions in plant cells (Geier 1980).

Another concern when using fixed and embedded tissues for histochemcial procedures is the removal of certain compounds during tissue processing. As a partial solution to this problem, fresh, unfixed tissue should be studied in conjunction with processed tissue wherever possible. Intact trichomes are often difficult to study in this way because the cuticular covering often impedes stain penetration.

More recent techniques like fluorescence microscopy and energy dispersive X-ray analysis have been useful in studying certain aspects of plant trichomes and it is anticipated that with further developments in techniques, additional information on the structure and composition of plant trichomes will be gained.

ACKNOWLEDGMENTS

We thank Melanie Howarth for her help in organizing the literature and proofing the manuscript, Doug Greenville for the use of photomicrographs for Figures 1 and 2 and for help with the plates, and the National Sciences and Engineering Research Council of Canada for financial support.

4. REFERENCES

Akers, C.P., J.A. Weybrew and R.C. Long, 1978, Ultrastructure of glandular trichomes of leaves of *Nicotiana tabacum* L., cv. Xanthi. Am. J. Bot. 65:82-92.

Amelunxen, F., 1964, Elektronenmikroskopische Untersuchungen an den Drüsenhaaren von *Mentha piperita* L. Planta med. 12:121-139.

Amelunxen, F., 1965, Elektronenmikroskopische Untersuchungen an den Drüsenschuppen von *Mentha piperita* L. Planta med. 13:457-473.

Amelunxen, F., T. Wahlig and H. Arbeiter, 1969, On the presence of essential oil in isolated glandular hairs and trichomes of *Mentha piperita* L. Z. Pflanzenphysiol. 61:68-72.

Andre, C. and A. Vercruysse, 1976, Histochemical study of the stalked glandular hairs of the female *Cannabis* plants, using fast blue salt. Planta med. 29:361-366.

Bancher, C. and J. Holzl, 1959, Über die Drüsenhaare von *Solanum tuberosum* Sorte *Sieglinde*. Protoplasma 50:356-369.

Barlow, P.W., 1975, The polytene nucleus of the giant hair cell of *Byronia* anthers. Protoplasma 83:339-349.

Barlow, P.W. and J.A. Sargent, 1975, The ultrastructure of the hair cells on the anther of *Bryonia dioica*. Protoplasma 83:351-364.

Beckman, C.H., W.C. Mueller and W.E. McHardy, 1972, The localization of stored phenols in plant hairs. Physiol. Plant Pathol. 2:69-74.

Bentley, N.J. and F.A. Wolf, 1945, Glandular leaf hairs of oriental tobacco. Bull. Torrey Bot. Club. 72:345-360.

Bonnot, E.J. and C. Hebant, 1970, Precisions sur la structure et let fonctionnement des mucigenes de *Polytrichum juniperinum* Willd. C.R. Hebd. Seances Acad. Sci. Ser. D. Sci. Nat. 271:53-55.

Charriere-Ladreix, Y., 1975, La secretion lipophile des bourgeons d'*Aesculus hippocastanum* L.: Modifications ultrastructurales des trichomes au cours du processus glandulaire. J. Microsc. (Paris) 24(1):75-90.

D'Amato, F., 1952, Polyploidy in the differentiation and function of tissues and cells in plants. Caryologia 4(3): 331-358.

Dayanandan, P. and P.B. Kaufman, 1976, Trichomes of *Cannabis sativa* L. (Cannabaceae). Am. J. Bot. 63(5):578-591.

Dell, B. and A.J. McComb, 1975, Glandular hairs, resin production, and habitat of *Newcastelia viscida* E. Pritzel. (Dicrastylidaceae) Aust. J. Bot. 23:373-390.

Dell, B. and A.J. McComb, 1977, Glandular hair formation and resin secretion in *Eremophila fraseri* F. Meull. (Myoporaceae) Protoplasma 92:71-86.

Dell, B. and A.J. McComb, 1978, Plant resins --- Their formation, secretion and possible functions. In Woodhouse, H.W. (Ed): Advances in Botanical Research pp. 217-316. London-New York: Academic Press.

Dro, A.S. and F.W. Hefendehl, 1973, Biogenese des ätherischen Öls von *Ocimum gratissimum*. Planta med. 24(4):353-366.

Fahn, A., 1979, Secretory Tissues in Plants. New York: Academic Press.
Figier, J., 1968, Etude infrastructurale et cytochimique des glandes petiolaires de *Mercurialis annua* L. Essai d'interpretation en rapport avec la secretion. C.R. Hebd. Seances Acad. Sci. Ser. D. Sci. Nat. 267:491-494.
Figier, J., 1969, Incorporation de glycine -- ^{3}H chez les glandes petiolaires de *Mercurialis annua* L. Etude radiographique en microscopie electronique. Planta 87:275-289.
Geier, T., 1980, PAS-positive reaction of phenolic inclusions in plant cell vacuoles. Histochemistry 65:167-171.
Hammond, C.T. and P.G. Mahlberg, 1978, Ultrastructural development of capitate glandular hairs of *Cannabis sativa* L. (Cannabaceae). Am. J. Bot. 65(2):140-151.
Hébant, C. and E.J. Bonnot, 1974, Histochemical studies on the mucilage-secreting hairs of the apex of the leafy gametophyte in some Polytrichaceous mosses. Z. Pflanzenphysiol. 72:213-219.
Heinrich, G., 1973, Über das ätherische Öl von *Monarda fistulosa* und den Einbau von markiertem CO_2 in dessen Komponenten. Planta med. 23(3):201-212.
Heinrich, G., 1977, Die Feinstruktur und das ätherische Öl eines Drüsenhaares von *Monarda fistulosa*. Biochem. Physiol. Pflanz. 171:17-24.
Horner, H.T. and N.R. Lersten, 1968, Development structure and function of secretory trichomes in *Psychotria bacteriophylla* (Rubiaceae). Am. J. Bot. 55(9):1089-1099.
Jakowska, S., 1949, The trichomes of *Physaria geyeri*, *Physaria australis* and *Lesquerella sherwoodii*. Development and morphology. Bull. Torrey Bot. Club 76(3):177-195.
Johnson, H.B., 1975, Plant pubescence: an ecological perspective. Bot. Rev. 41:233-258.
Juniper, B.E., A.J. Gilchrist and R.J. Robbins, 1977, Some features of secretory systems in plants. Histochem. J. 9:659-680.
Kreitner, G.L. and E.L. Sorensen, 1979, Glandular secretory system of alfalfa species. Crop Sci. 19:499-502.
Kristen, U., 1976, Die Morphologie der Schleimsekretion in Fruchtknoten von *Aptenia cordifolia*. Protoplasma 89:221-233.
Landre, P., 1976, Evolution of nuclear DNA content in secretory trichome cells of *Solanum nigrum* L. during their formation. Caryolgia 29(2):235-245.
Lassanyi, Z., G. Stieber and E. Tyihak, 1978, Investigations into the volatile oil secretory system of *Chamomile anthodium*. Part I. The histochemical analysis of glandular hairs. Herba Hung. 17(2):31-42.

Lersten, N.R. and J.D. Curtis, 1977, Preliminary report of outer wall helices in trichomes of certain dicots. Can. J. Bot. 55(2):128-132.

Lersten, N.R., and H.T. Horner, Jr., 1967, Development and structure of bacterial leaf nodules in *Psychotria bacteriophila* Val. (Rubiaceae). J. Bact. 94:2027-2036.

Levin, D.A., 1973, The role of trichomes in plant defense. Q. Rev. Biol. 48(1):3-15.

Luttge, U., 1971, Structure and function of plant glands. Ann. Rev. Plant Physiol. 22:23-44.

Lyshede, O.B., 1976, Structure and function of trichomes in *Spartocytisus filipes*. Bot. Notiser 129:395-404.

Nielson, A.J. and W.P. Griffith, 1978, Tissue fixation and staining with osmium textroxide: The role of phenolic compounds. J. Histochem. Cytochem. 26:138-140.

O'Brien, T.P., 1972, The cytology of cell wall formation in some eukaryotic cells. Bot. Rev. 38:89-118.

Olesen, P., 1979, Ultrastructural observations on the cuticular envelope in salt glands of *Frankenia pauciflora*. Protoplasma. 99:1-9.

Peterson, R.L. and L.S. Kott, 1974, The sorus of *Polypodium virginianum:* some aspects of the development and structure of paraphyses and sporangia. Can. J. Bot. 52:2283-2288.

Pizzolato, P., 1964, Histochemical recognition of calicum oxalate. J. Histochem. Cytochem. 12:333-336.

Rachmilevitz, T. and D.M. Joel, 1976, Ultrastructure of the calyx glands of *Plumbago capensis* thunb. in relation to the process of secretion. Isr. J. Bot. 25:127-139.

Rougier, M., 1966, Differenciation cellulaire et presence de mucopolysaccharides dans les squamules de l'apex d'*Elodea canadensis*. C.R. Soc. Biol. 160:2182-2183.

Rougier, M., 1972, Etudie cytochimique des squamules d'*Elodea canadensis*. Protoplasma 74:113-131.

Sakai, W.S., 1974, Scanning electron microscopy and energy dispersive X-ray analysis of chalk secreting leaf glands of *Plumbago capensis*. Am. J. Bot. 61(1):94-99.

Schnepf, E., 1968, Zur Feinstruktur der schleimsezernierenden Drüsenhaare auf der Ochrea von *Rumex* und *Rheum*. Planta (Berl.) 79:22-34.

Schnepf, E., 1974, Gland Cells. In Robards, A.W. (Ed.): Dynamic aspects of plant ultrastructure. pp. 331-357. London: McGraw Hill.

Schnepf, E. and J. Busch, 1976, Morphology and kinetics of slime secretion in glands of *Mimulus tilingii*. Z. Pflanzenphysiol. 79:62-71.

Schnepf, E. and A. Klasova, 1972, Zur Feinstruktur von Öl und Flavon-Drüsen. Ber. Dtsch. Bot. Ges. 85(5/6):249-258.

Stahl, E., 1953, Untersuchungen an den Drüsenhaaren der Schafgarbe *(Achillea millefolium* L.) Z. Bot. 41:123-146.
Stahl, E., 1957, Über Vorgänge in den Drüsenhaaren der Schafgarbe. Z. Bot. 45:297-315.
Stahl, E., 1969, Thin-layer chromatography. Berlin-Heidelberg. New York: Springer-Verlag.
Szentpetery, R.G., A. Kovacs and S. Sarkany, 1969, Volatile oil excretion of the differentiating epidermis on the developing leaf of *Valeriana collina* Wallr. Acta agron. hung. 18(3-4):287-296.
Tiwari, S.C., 1978, Some unusual features of floral trichomes and nectaries in *Grewia subinaequalis.* Acta Bot. Indica 6(1):81-86.
Tschermak-Woess, E. and G. Hasitschka, 1953, Veränderungen der Kernstruktur während der Endomitose, rhythmisches Kernwachstum und verschiedenes Heterochromatin bei Angiospermen. Chromosoma 5:574-614.
Tschermak-Woess, E. and G. Hasitschka, 1954, Über die endomitotische Polyploidisierung im Zuge der Differenzierung von Trichomen und Trichozyten bei Angiospermen. Österr. Bot. Z. 101:79-117.
Turala, K., 1960, Endomitotical processes during the differentiation of the anthers' hairs of *Cucumis sativus* L. Acta Biol. Cracov. Ser. Botanica 3(1):1-13.
Turala, K., 1965, Mechanizmy cytologicznew. Toku roznicowania wloskow u *Echinocystis lobata* (Cytological processes during the differentiation of the hairs of *Echinocystis lobata*). Acta Biol. Cracov. Ser. Botanica 5(2):151-169.
Unzelman, J.M. and P.L. Healey, 1972, Development and histochemistry of nuclear crystals in the secretory trichome of *Pharbitis nil.* J. Ultrastruct. Res. 39:301-309.
Unzelman, J.M. and P.L. Healey, 1974, Development, structure and occurrence of secretory trichomes of *Pharbitis.* Protoplasma 80:285-303.
Uphof, J.C., 1962, Plant hairs. In Zimmerman, W. and Ozenda, P.G. (Eds.) Handbuch der Pflanzenanatomie. Band IV Teil 5.
Vermeer, J. and R.L. Peterson, 1979a, Glandular trichomes on the inflorescence of *Cyrysanthemum morifolium* cv. *Dramatic* (Compositae). I. Development and morphology. Can. J. Bot. 57:705-713.
Vermeer. J. and R.L. Peterson, 1979b, Glandular trichomes on the inflorescence of *Chrysanthemum morifolium* cv. *Dramatic* (Compositae). II. Ultrastructure and histochemistry. Can. J. Bot. 57:714-729.
Wanstall, P.J., 1950, Mucilage hairs in *Polytrichum.* Trans. Br. bryol. Soc. 1:349-352.
Wardlaw, C.W., 1970, Cellular differentiation in plants and other essays. Manchester: Manchester University Press.

Weber, U. and J. Deufel, 1951, Zur Cytologie der Drüsenhaare von
    Achillea millefolium. Arch. Pharm. (Weinheim) 284(56):
    318-323.
Wollenweber, E. and E. Schnepf, 1970, Comparative studies on
    the flavonoid excretions of farina and oil excreting glands
    of primroses and the fine structure of the gland cells.
    Z. Pflanzenphysiol. 62:216-227.

CELLULAR BASIS OF TRICHOME SECRETION

William W. Thomson and Patrick L. Healey

Department of Botany and Plant Sciences
University of California, Riverside

and

Department of Developmental and Cell Biology
University of California, Irvine

ABSTRACT

Three types of secretory trichomes were analyzed to determine correlations between their ultrastructure and cytochemistry, and the products they secrete. These included salt secreting glands of Atriplex *and other genera, mucilage secreting trichomes of* Pharbitis, *and the lipid secreting glands of* Phacelia. *It is clear that each shows complex mechanisms which do not fit conveniently into any previously proposed classification system. More precise localization of the sites of synthesis and transport are required before final conclusions can be reached regarding any commonalities which might exist among mechanisms used by trichomes which secrete a wide variety of substances.*

1. INTRODUCTION

Secretion is a fundamental process characteristic of all living cells. Mediated by a membrane or membrane system, it requires, in eukaryotic cells, a degree of metabolic regulation which is concomitantly dependent on membrane structure and organization, the spatial distribution of cellular compartments, as well as the functional, structural and developmental inter-relationships between membranes, organelles and compartments. Many trichomes are highly differentiated for the process of secretion, and, thus, they offer avenues to mount integrated investigations on the structural and developmental bases for this process. These extend from the level of membrane structure and organization to compartmentalization and the three-dimensional organization of the cells.

Over the past decade or so some useful mechanistic models have been proposed for different modes of secretion and here we are particularly referring to the categories presented by Schnepf (1969) and Fahn (1979). This includes elimination from the cytoplasm of the cell by transport or diffusion across the plasmalemma or tonoplast (eccrine); vesicular transport from the cytoplasm to the plasmalemma or tonoplast (granulocrine); elimination of material from the plant through the death of cells (holocrine). Other attempts to generalize about secretion and secretory trichomes have been to identify and categorize glands by predominate cytological features such as plastids or endoplasmic reticulum. The implications here are that such structures are of primary importance in any or all such processess such as synthesis, processing, and secretion of the final secretory product (Lüttge and Schnepf 1976). Similarly, there have been some suggestions that cytological and cell organizational features can be correlated with the nature of the secreted material such as proteins, carbohydrates or lipophilic material (Lüttge and Schnepf 1976).

In this chapter we would like to discuss some of these concepts in relation to secretory trichomes that we have studied. These include salt glands, the trichomes of *Pharbiti* which secrete a protein-carbohydrate mucilage, and the trichomes of *Phacelia* which secrete a phenolic, dermatotoxic substance.

2. RESULTS

From an anatomical and ultrastructural point of view there are three different types of salt glands (Thomson 1975). One form is illustrated by the trichomes found on leaves of *Atriplex* (Fig. 1) in which there are one or more stalk cells terminated by a large, vacuolated bladder cell. In these glands the actual secretion onto the surface of the leaf appears to be an integral function of the development of the bladder cells (Thomson and

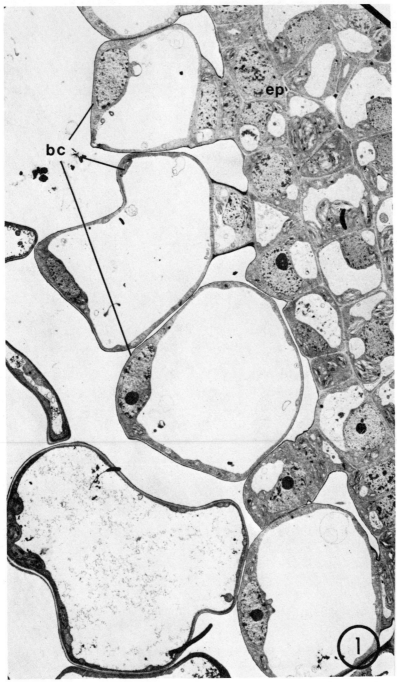

Fig. 1. A low magnification micrograph of trichomes projecting from the epidermis (ep) of a leaf of *Atriplex*. The large, terminal bladder cells (bc) are well developed.

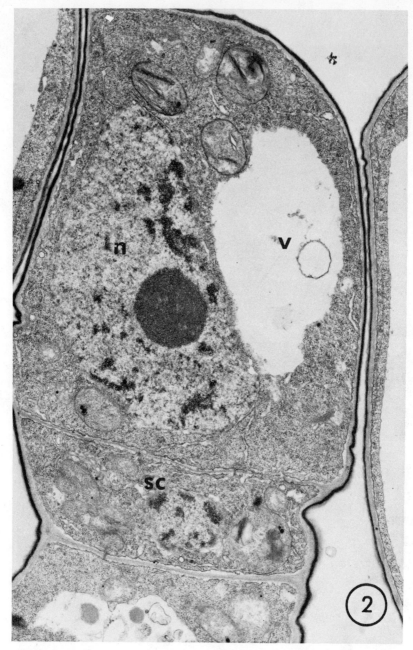

Fig. 2. A developing *Atriplex* trichome. The enlarging bladder cell above the stalk cell (sc) is characterized by a large nucleus (n) and a dense cytoplasm containing numerous ribosomes and other organelles. At this stage the vacuoule of the bladder cell (v) is expanding concomitantly with the expansion of the cell. X 16,000.

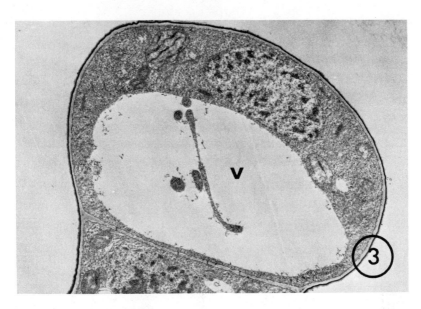

Fig. 3. This micrograph further illustrates that the expansion of the bladder cell and the vacuole (v) occurs concomitantly. X 8,3000.

Platt-Aloia 1979). That is, most evidence indicates that salts accumulate in the central vaucole of the developing bladder cell and with accumulation there is a concomitant expansion of the vacuole and bladder cell (Figs. 1, 2, 3). This accumulation and expansion continues until the bladder cell ruptures, releasing the salts to the surface of the leaves (See Luttge 1971, Thomson 1975, Hill and Hill 1976 for reviews). This process of secretion can be considered holocrine since it involves the death of the bladder cell. This interpretation may, however, be somewhat simplistic, since the underlying mechanism involved is salt accumulation in the vacuole which, presumably, involves a membrane transport function. In other words the primary mechanism involved would be, in part, an eccrine mode of secretion.

Most physiological evidence indicates that accumulation of salts in the cell is probably active but there is little direct information available on the mechanism(s) involved in the accumulation of salt within the vacuole (See Lüttge 1971, Thomson 1975, Hill and Hill 1976 for reviews). Ultrastructural studies have shown that early in development there is a transition in ribosomes from being predominantly free to clustered, and there is a significant level of ribosomes, RER, and apparently active dictyosomes even in greatly expanded bladder cells (Thomson and Platt-Aloia, 1979). Nevertheless, it is difficult to differentiate a special role for these cellular elements in salt accumulation against their probable function in other

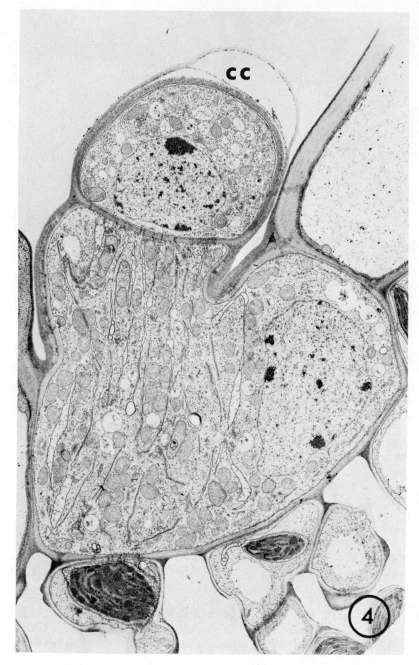

Fig. 4. This micrograph depicts the two-celled salt gland of the grass, *Cynondon*. Note the large, flask-shaped, basal cell with the terminal cap cell projected above the epidermis by the neck of the basal cell. X 5,750.

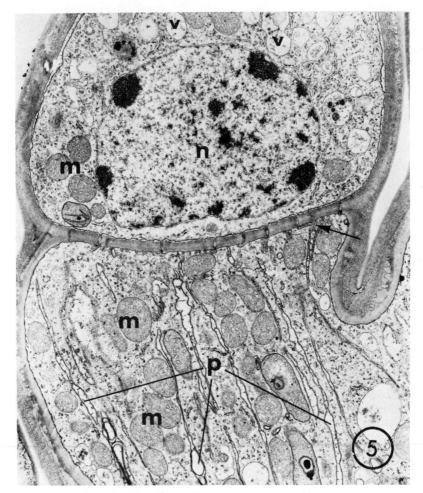

Fig. 5. This micrograph illustrates three major aspects of the salt glands in monocotyledons: First, the numerous partitioning membrances (p) and associated mitochondria (m) in the basal cell; second, the connection of the partitioning membranes (arrow) with plasmalemma along the wall between the basal and cap cell; third, the large nucleus and numerous organelles, ribosomes, and small vacuoles in the cap cell. X 11,500.

processes associated with the expansion of the bladder cell such as continuous wall formation and expansion and enlargement of the plasmalemma and tonoplast membranes.

In most monocotyledons, the salt glands are relatively simple, at least in cell number, and generally consist of two cells – a large basal cell which extends across the plane of the epidermal cells with an expanded portion embedded in, and associated with, the underlying mesophyll cells, and a smaller, outer cap cell which extends somewhat above from the surface of the leaf (Fig. 4). The most striking features of the basal cell are the numerous partitioning membranes (Fig. 5, p). These membranes are closely associated with mitochondria, and this complex of mitochondria and partitioning membranes could represent a functional unit (Copeland 1966). The partitioning membranes are invaginations of the plasmalemma which extend into the basal cell from the cross wall between the cap and basal cell (Fig. 5, arrow). These invaginations extend nearly to the base of the basal cell, but in this region of the cell no connections with the plasmalemma can be observed. Thus, the channel between the membranes is extracellular, closed at the basal terminus, but open to the wall or apoplast at the juncture with the cap cell. That is, the channel is open in the direction of salt secretion from the glands. Earlier, Levering and Thomson (1971) proposed that the partitioning membranes and the extracellular channels probably had a significant role in the secretory process. They suggested that a coupled solute-water transport may be involved similar to that which has been proposed for many animal tissues with similar membrane geometry. That is, solutes are probably transported across the partitioning membranes into the channels with osmotic flow of water, the salt solution flowing by diffusion along and out of the channels. Eventually, it was envisioned that the salt solution would be emitted to the surface of the leaves through apparent pores in the cuticle which covers the cap cell. Dilations of the channels under salt-loaded, secreting conditions tend to support this conclusion (Levering and Thomson 1972). However, this model is extremely simplistic for at least two reasons. First, it does not account for, or establish, a functional role for the cap cell in the secretion process. Since the cap cell is positioned between the basal cell and the point where salts are emitted to the surface of the leaf, it is difficult to imagine that it has no role in the secretion process. Further, the cap cells contain sufficient cell organelles such as mitochondria, endoplasmic reticulum, ribosomes and small vacuoles (Fig. 5) to indicate that they have more than a minimal metabolic activity and it seems reasonable to suspect that they do function in the secretory process. Second, the extracellular channel defined by the invaginated partitioning membranes opens to the wall between the basal and cap cell. However, this wall has an apoplastic continuum with the wall of the adjacent and underlying mesophyll, as well as the lateral and outer wall of the cap cells. For the salts to diffuse vectorially outwards to the cuticle, the continuum with the leaf apoplast must be restricted to prevent backflow into the tissue. However, no cuticularization or suberinization of these walls have been noted. Thus, the vectorial pathway of movement of salts from the basal cell to the surface of the leaf has not been completely determined.

The salt glands which occur on plants such as *Tamarix Frankenia, Limonium* and many of the mangroves tend to be multicellular (Thomson 1975). Many of these are epidermal structures;

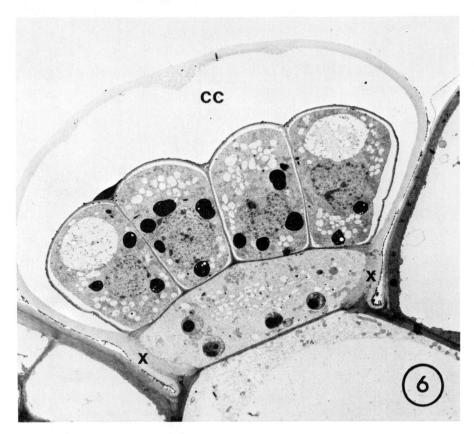

Fig. 6. A large, multicellular salt gland on the leaf of the mangrove, *Avicennia* Note large collecting compartment (cc) between the secretory cells and the expanded cuticule and that the cuticle extends along the lateral walls of the basal cell (x). The micrograph has been rotated ninety degrees to the left for illustration purposes. X 2,000.

however, some, such as those on the abaxial surface of *Avicennia* leaves, are truly trichomes (Fig. 6). In these glands a large, apparently expandable, "collecting chamber," occurs between the secretory cells and the covering cuticle (Fig. 6, cc) and there are small pores in this cuticle in the glands of monocots (Fig. 4, cc). Another characteristic feature of several of these glands, is that the lateral, anticlinal walls of the basal cells are completely cutinized (Fig. 6, 7, x). Thus, apoplastic continuity of the walls of the outer secretory cells with those of the underlying mesophyll or adjacent epidermal cells is restricted. Although quantitative studies have not been done, inactive (non-secreting) gland cells of *Tamarix* and *Frankenia* have few ribosomes, a limited amount of endoplasmic reticulum

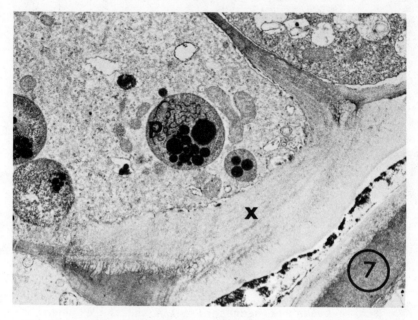

Fig. 7. An enlarged micrograph which illustrates that the lateral wall of the *Avicennia* salt gland is completely cuticularized. X 8,500.

inactive (non-secreting) gland cells of *Tamarix* and *Frankenia* have few ribosomes, a limited amount of endoplasmic reticulum (ER), and lack or have few vacuoles. In contrast, in active (secreting) glands there are numerous ribosomes, a noticeable increase in ER and numerous microvacuoles (Thomson et al. 1969, Campbell and Thomson 1976a, b).

The association of many of the microvacuoles with the plasmalemma of the secretory cells, and accumulations of electron dense material within the microvacuoles of rubidium-treated material led to the suggestion that salts are accumulated in the microvacuoles and released into the walls of the glands when the microvacuoles fuse with the plasmalemma (Thomson and Liu 1967, Thomson et al. 1969, Campbell and Thomsom 1976a,b). Also, since studies on enzymes from halophytic species indicate that they are as sensitive to salts as those from salt-sensitive plants (Flowers 1972, 1975, Greenway and Osmond 1972, Hall and Flowers 1973), compartmentalization of the salts in the vacuoles of the cortical cells and microvacuoles of the glands has been proposed as the mechanism for protection of these enzyme systems (Flowers 1975).

However, except for the presence of electron dense material in the microvacuoles of rubidium-treated material, which was absent in controls (Thomson et al. 1969), little other evidence is available to support or negate the suggestion that salts are accumulated in the microvacuoles. Thus, a clear delineation of the role of the microvacuoles in salt secretion remains unresolved. For example, Hill and Hill (1976) have argued that vesicles may be simple inward inflations of the plasmalemma and Lüttge (1971) has questioned whether the loading of the microvacuoles may occur in parallel with another mechanism of transport out across the plasmalemma. It is obvious that a determination of the site(s) of salt accumulation in the glands is needed.

There is ample physiological evidence that salt secretion is an active process and requires the input of energy by the cells (Thomson 1975, Hill and Hill 1976). In considering the different types of salt glands, we would hypothesize that a membrane transport process, i.e. an eccrine form of secretion, is the primary mechanism involved in *Atriplex* through the accumulation of salt in the vacuole across the tonoplast membrane, in the two-celled glands of monocots, by transport across the "partitioning membranes" into the extracellular channels, and in the multicellular salt glands, across the membrane of the microvacuoles (or their source membrane) and/or the plasmalemma of the main secretory cells.

In regard to the actual secretion or emission of salts to the surface of the leaf, we would also like to suggest that a common mechanism is involved in all three types of salt glands. Although secretion appears to be an active process, most measurements indicate that the osmotic concentration of the secreted fluid is equal to or slightly higher than that of the leaf or xylem sap (Arisz et al. 1955, Atkinson et al. 1967, Scholander et al. 1962, Scholander et al. 1966). Why is the secreted fluid not significantly more concentrated than that of the leaf? If the process is active, it should be against an electrochemcial potential. Previously, we have suggested that a standing-gradient flow system is the operative mechanism of emission of salts to the surface of the leaves in the two-celled glands of the grasses (Levering and Thomson 1971, Thomson 1975). In the operation of the standing-gradient flow system, we are extrapolating from the model proposed by Diamond and Bossert (1967) in which solutes are pumped actively into the closed end of the channels forming a localized gradient followed by a passive flow of water. As the solute diffuses towards the open end of the channel, the fluid becomes progressively less hypertonic as more water enters along the channel. We are suggesting that this process probably functions, at least in a general sense, in the secretion of salts from the other glands as well. In the standing-gradient flow system, a compartment is required for the coupling of solute transport with the osmotic movement of water (Diamond and Bossert 1967, Diamond 1971).

For the multicellular glands we proposed that the salts are transported into the cavity between the outer secretory cells and the overlying cuticle - the "collecting chamber." With this transport, water flows osmotically into the chamber resulting in a decrease in the osmotic concentrations of the eventually secreted fluid. With increasing accumulation of fluid, the chamber enlarges by distention of the overlying cuticle. When sufficient pressure develops the fluid is released through the cuticular pores (Thomson 1975).

The tight association of the plasmalemma with the cuticle along the basal cells of the glands apparently prevents a pressurized backflow into the leaf (Thomson and Liu 1967). Although there is little direct evidence as yet, the vacuole of *Atriplex* may function in a similar manner, i.e. solute accumulation followed by water flow leading to expansion of the bladder cell until it bursts, releasing the salt.

As compared with salt glands where the secretion consists of mineral elements and water both of which are delivered to the leaves by the transpiration stream and to the glands via either or both the apoplastic and symplastic pathways, trichomes which secrete predominately organic materials must be viewed somewhat differently when trying to understand the overall process of secretion. This is because it is generally assumed but seldom established that the secreted product is synthesized or at least significantly modified by the gland cells. This is not to say that some aspects of the secretory mechanism may be similar to what occurs in salt glands, but for a complete understanding of the secretory processes it seems requisite to consider the sites of synthesis and possible processing as well as the vectorial pathway for the movement of the material from the site of synthesis to the point(s) of secretion.

The mature, secreting trichomes of *Pharbitis* consist of four cap or secretory cells which are attached to a foot cell which projects the bulbous head of the trichome above the epidermis. The anticlinal walls of the stalk cells are apparently completely cutinized and a thin cuticule extends over the cap cells which expands away from the walls during secretion. Unzelman and Healey (1974) determined that the secreted, mucilagenous material from these glands was a complex of protein and carbohydrate using cytochemical techniques. When the glands are mature and actively secreting, the cap cells are characterized by an extensive, lace-like network of rough endoplasmic reticulum. The RER surrounds large, apparently golgi derived storage vesicles which contain both protein and carbohydrate. Although coated vesicles are associated with the plasmalemma and golgi vesicles, in regard to secretion what is striking is the numerous attachments of the RER with the plasmalemma. Unzelman and Healey (1974) suggested that secretions may thus be transported through the cell and emitted to the wall via the continuity of this cytoplasmic membrane system with the plasmalemma. This suggestion deviates from the somewhat more popular view that secretion is mediated by golgi vesicles and exocytosis. However, as O'Brien (1972), Chrispeels (1976) and Rothman (1975) have pointed out, the evidence for the ubiquity of vesicular mediated secretion, particularly in plants, is far from secure. The possibility that the transfer of cytoplasmicly synthesized products to outside of the cells by way of ER connections with the plasmalemma has more than syntactic implications.

The ultrastructure of the secretory trichomes of *Phacelia* suggest that the mechanism of secretion, although similar in some aspects, is somewhat different to that which occurs in *Pharbitis* (Trautner and Healey 1980). This may be related to the fact that secretory products are different in that they consist primarily of lipid, carbohydrate, phenylated quinones and flavonoids (see Kelsey et al. 1983) The most striking aspects of the ultrastructure of the secretory cells is a massive proliferation of smooth endoplasmic reticulum near the wall of the glands and a multiple array of membraneous cisternae which are partially encapsulated around membrane bound vesicles. As in other glands such predominant features and their spatial relationships encourages the conclusion that they have an integrated and fundamental role in the secretory process. The secreted material is extremely electron dense and deposits of similar density occur in the walls, between the plasmalemma and the wall, in the SER and dilated vesicles of the SER, and in the cisternae surrounding the membrane bound vesicles. Making the reasonable assumption that these deposits represent material eventually to be secreted, its pattern of distribution in these associated membrane systems lends credence to the view that the entire complex is involved in the synthesis, and vectorial transfer of the secretion to the surface of the cells. Observed fusion of the SER and the dilated SER vesicles with the plasmalemma also adds to the view that this membrane system is a primary transfer element in the secretory process. In *Phacelia* this transfer system appears to be SER which in *Pharbitis* consists of RER. This may be related to the secreted material consisting in part of lipids (and possibly phenolics) and the observation that SER is commonly an abudant feature in other plant glands that secrete lipid or lipophilic material (Lüttge and Schnepf 1976, Fahn 1979).

3. DISCUSSION

Since we have integrated many of our views and interpretations with the presentation of our observations, we would like to summarize where we see studies on the structure and function of secretorty trichomes to be.

First, these types of trichomes can be recognized as highly specialized for secretion, and products secreted represent almost every major class of compounds which occur in plants (except possibly nucleic acids) as well as mineral elements. Thus, the trichomes have attracted the attention of a broad spectrum of investigators from ecologists, studying the role of glands in the adaptation of plants to particular environments, to physiologists who view the glands as a means to analyze and determine the role in secretion of cellular compartmentation, membrane interrelationships, and spatial and developmental interrelationships of cell organization as a function of cellular morphogenesis.

Second, to date a fair number of secretory trichomes have been studied in some detail and in this paper we have enlarged on a few we have studied. As pointed out in the introduction, some general models and classifications of trichomes have been presented in the literature, either based on various postulated mechanisms of secretion from the cell or dominant organizational features. Again, these views have uses in the analysis of working hypotheses, but we are reluctant at this point to promulgate these views strongly. Our reasons are as follows: 1) We are not convinced that any one postulated mechanism can explain the general secretory process even for a single type of gland. For example, secretion from salt glands at some point may involve a primary transport of ions across membrances (eccrine secretion) but this process can not account completely for the secretion of salts from the glands to the surface of the leaves; 2) Identifying glands as to type strictly as to the products secreted and to apparently a related common organizational feature may also be misleading. This is because the nature of the secretion is often not known with any certainty and when even partially characterized it often consists of a variety of compounds. To assume that the same mechanisms exists for the vecotorial transport of these different compounds to the plasmalemma and their subsequent release to the extracellular milieu, to us, seems limiting. Even though there is abundant ER in both the glands of *Pharbitis* and *Phacelia* as well as fusion sites of the ER with the plasmalemma, we can not be certain, for example, in *Phacelia*, that the secretion of all the compounds, lipids, phenolic, and carbohydrates occurs via this system.

Third, we have suggested at least for *Pharbitis* and *Phacelia*, that the ER may have a primary vectorial role in the movement of secretory material from within to the outside of the cell via continuities with the plasmalemma. However, this suggestion further implies that the actual extrusion of the secreted material into the extracellular continuum of the ER cisternae may occur vectorially across the ER at the site of synthesis within the cell.

Finally, although a considerable body of literature has appeared on secretion by trichomes, we are convinced that the field is in its infancy. Considerable gains in our understanding could be achieved, and the tools are increasingly available, if the products of secretion were more critically characterized. This could be accomplished with the application of careful autoradiographic studies, at both the light and electron microscope level, to achieve a more exact localization of secretory products within the cells as well as biochemical evidence for the site(s) of this synthesis.

At the ultrastructural level, the amounts, positions, and particular spatial relationships of cell organelles, compartments and membranes, seem to be a reasonable basis for the development of a framework of ideas as to functional interrelationships involved. However, in densely cytoplasmic cells, as most gland cells are, there is a definite possibility that some apparent associations are merely accidental juxtapositions. For example, plastids occur in almost, if not all, living cells including gland cells, and they are recognized to have a multiplicity of function (Thomson and Whatley 1980). The question is, however, whether they have a primary or a secondary role in synthesis and secretion in these glands. Resolution of these types of questions will be difficult but more valid considerations could be advanced as to the significance of interrelationships if quantitative and three dimensional determination were made using morphometric and stereological techniques. This would seem particularly valuable if comparisons were made between secreting and non-secreting glands.

ACKNOWLEDGEMENTS

We acknowledge the support of the NSF Research grants, PCM74-19987 and PCM-80003779, to WWT in this research.

4. REFERENCES

Arisz, W.H., Camhuis, I.J., Heikens, H., and van Tooren, A.J., 1955, The secretion of the salt glands of *Limonium latifolium* Ktze. Acta Bot. Neerl. 4:322-338.

Atkinson, M.R., Findlay, C.P., Hope, A.B., Pitman, M.C., Saddler, H.D.W., and West, K.R., 1967, Salt regulation in the mangroves *Rhizophora mucronata* Lam. and *Ageialitis annulata* R. Br. Aust. J. Biol. Sci. 20:589-599.

Campbell, N., and Thomson, W.W., 1976a, The ultrastructure of *Frankenia* salt glands. Ann. Bot. 40:681-686.

Campbell, N., and Thomson, W.W., 1976b, The ultrastructural basis of chloride tolerance in the leaf of *Frankenia*. Ann. Bot. 40:687-693.

Chrispeels, M.J., 1976, Biosynthesis, intracellular transport, and secretion of extracellular macromolecules. Ann. Rev. Plant Physiol. 27:19-38.

Copeland, E., 1966, Salt transport organelle in *Artemia satenis*. Science 151:470-471.

Diamond, J., 1971, Water-solute coupling and ion selectivity in epithelia. Phil. Trans. Roy. Soc. Lond. B262:141-151.

Diamond, J.M., and Bossert, W.H., 1967, Standing-gradient osmotic flow. A mechanism for coupling water and solute transport in epithelia. Jour. Gen. Physiol. 50:2061-2083.

Fahn, A., 1979, Secretory tissues in plants. Academic Press, 302, London, New York, San Francisco.

Flowers, T.J., 1972, The effect of sodium chloride on enzyme activity from four halophyte species of Chenopodiaceae. Phytochem. 11:1881-1886.

Flowers, T.J., 1975, Halophytes. *In:* Baker, D.A., Hall, J.L. (eds.): Ion Transport in Plant Cells and Tissues, p. 309-334, North Holland Pub. Co., Amsterdan and London.

Greenway, H., and Osmond, C.B., 1972, Salt responses of enzymes from species differing in salt tolerance. Plant Physiol. 49:256-259.

Hall, J.L. and Flowers, T.F., 1973, The effect of salt on protein synthesis in the halophyte *Suaeda maritima*. Planta 110:361-368.

Hill, A.E., and Hill, B.S., 1976, Mineral ions. *In:* Lüttge U., and Pitman, M.G. (eds.): Transport in Plants II: Part B Tissues and Organs, p. 225-243, Springer-Verlag, Berlin, Heidelberg, New York.

Kelsey, R., Reynolds, G, and Rodriguez, E., 1983, The chemistry
of biologically active constituents secreted and stored
in plant glandular trichomes, see Chapter 8, this book.

Levering, C.A., and Thomson, W.W., 1971, The ultrastructure of
the salt gland of *Spartina foliosa*. Planta 97:183-196.

Levering, C.A., and Thomson, W.W., 1972, Studies of the
ultrastructure and mechanism of secretion of the salt gland
of the grass *Spartina*. Proc. 30th Electron MIcroscopy Soc.
Amer., 22-223.

Lüttge, U., 1971, Structure and function of plant glands. Ann.
Rev. Plant Physiol. 22:23-44.

Lüttge, U., Schnepf, E., 1976, Organic substances. *In:* Lüttge,
U., and Pitman, M.G. (eds.), Transport in Plants II:
Part B. Tissues and Organs, pp. 244-277, Academic Press,
Berlin, Heidelberg, New York.

O'Brien, T.P., 1972, The cytology of cell-wall formation in
some eucaryotic cells. Bot. Rev. 38:87-118.

Rothman, S.S., 1975, Protein transport by the pancreas. Science
190:747-753.

Schnepf, E., 1969, Sekretion und exkretion bei pflanzen.
Protoplasmatologia Vol. VIII/8. Wein - New York, Springer-
Verlag, p. 181.

Scholander, P.F., Hammel, H.T., Hemmingsen, E.A., and Garry, W.,
1962, Salt balance in mangrovres. Plant Physiol. 37:722-729.

Scholander, P.F., Bradstreet, E.D., Hammel, H.T., and Hemmingsen,
E.A., 1966, Sap concentration in halophytes and some other
plants. Plant Physiol. 41:529-532.

Thomson, W.W., 1975, The structure and function of salt glands.
In: Poljakoff-Mayber, A., and Gale, J. (eds.), Plants in
Saline Environments, Analysis and Synthesis 15:118-146,
Springer-Verlag, New York.

Thomson, W.W., and Liu, L.L., 1967, Ultrastructural features of
the salt glands of *Tamarix aphylla* L. Planta, 73:201-220.

Thomson, W.W., and Platt-Aloia, K., 1979, Ultrastructural transi-
tions associated with the development of the bladder cells
of the trichomes of *Atriplex*. Cytobios. 25:105-114.

Thomson, W.W., Berry, W.L., and Liu, L.L., 1969, Localization
and secretion of salt by the salt glands of *Tamarix aphylla*.
Proc. Nat. Acad, Sci. 63:310-317.

Thomson, W.W., and Whatley, J.M., 1980, Development of nongreen
plastids. Ann. Rev. Plant Physiol. 31:375-394.

Trautner, R., and Healey, P.L., 1980, personal communication.

Unzelman, J.M., and Healey, P.L., 1974, Development, structure
and occurrence of secretory trichomes of *Pharbitis*.
Protoplasma 80:285-303.

ECOLOGY AND ECOPHYSIOLOGY OF LEAF PUBESCENCE IN NORTH AMERICAN DESERT PLANTS

James Ehleringer

Department of Biology, University of Utah

Salt Lake City, UT 84112

Abstract

The ecophysiological effects of leaf trichomes on surface spectral characteristics and boundary layer thickness were investigated for a number of Sonoran Desert plants. Dense trichome layers substantially increase leaf reflectance for all wavelengths of solar radiation between 400-3000 nm. Leaf absorptance to total incident solar radiation may be decreased by a factor of three when compared to the leaf absorptance of glabrous leaves. These changes in leaf absorptance have an effect on leaf temperature, photosynthetic rate, and transpiration rate. Leaf boundary layers are only slightly increased by dense tomentum in the desert species studied. These ecophysiological effects of trichomes on leaf activity are discussed with respect to life form and to adaptation to arid land habitats.

1. INTRODUCTION

During the past century a correlation has been established between the presence of leaf trichomes and aridity in higher plants. Haberlandt (1884), Schimper (1903) and Warming (1909) have all pointed out that leaf pubescence is a common feature of plants from Mediterranean climates and steppe, desert, and alpine habitats. These early investigators also reported that, when an individual species extended over a broad environmental gradient, in mesic habitats the species had glabrous or glabrate leaves whereas in xeric habitats the leaves were pubescent. Moreover, in many habitats the leaves of some species were glabrate during the moist period of the year and pubescent during dry periods. Johnson (1968) has studied leaf pubescence in four communities (sandy beach, old field, oak forest, and red maple swamp) which represent an increasing moisture gradient in eastern North America. He found that although the incidence of pubescence was similar in each community (70-80% of the species), the pubescence was much more dense in the drier environments.

The adaptive value of leaf hairs is generally thought to be related to water economy of the plant either through (1) an increased reflection of solar radiation which reduces leaf temperature and thus transpiration rate, or (2) by increasing the thickness of the boundary layer (layer of still air over leaf through which water must diffuse) thereby reducing transpiration rate (Haberlandt 1884; Shull 1929; Wooley 1964; Wuenscher 1970; Ehleringer and Mooney 1978). Leaf pubescence does increase leaf reflectance in many species (Shull 1929; Billings and Morris 1951; Pearman 1966; Sinclair and Thomas 1970; Ehleringer et al. 1976; Ehleringer and Björkman 1978; Ehleringer 1981a), but exceptions occur (Shull 1929; Gausman and Cardenas 1969, 1973; Wuenscher 1970). The effects of pubescence on reflectance most likely depend on the density and thickness of the indumentum. The data relating to the leaf boundary layer effect are in conflict (Haberlandt 1884; Wiegand 1910; Sayre 1920; Wooley 1964; Wuenscher 1970; Parkhurst 1976), even when only studies of the same species are considered (Sayre 1920; Wuenscher 1970; Parkhurst 1976).

Recent studies suggest that leaf pubescence may play a role in reducing herbivory by serving as a physical barrier to animal penetration or by emitting toxic or repellent compounds (Levin 1973; Johnson 1975). Leaf hairs could serve both antiherbivore and water economy functions, but no data are available to resolve this point.

Increases in leaf pubescence are common along aridity gradients in the southwestern United States and northwestern Mexico (Shreve and Wiggins 1964; Ehleringer 1980). The pubescence is of various types (single cell, multicellular, branched, stellate, etc.) and occurs in almost all of the higher plant families along these aridity gradients. Such ecological gradients are ideal for studying the effects and ecological significance of leaf pubescence. Since the variations in leaf pubescence are strongly correlated with aridity, and apparently not with major distributional differences in animal taxa, the pubescence gradient is more likely a response to abiotic and not biotic components.

In this paper, I would like to briefly review the physiological and ecological significance of leaf pubescence to plants in arid environments.

2. DISCUSSION

2.1 *Pubescence and Aridity Gradients*

The increase in leaf pubescence along an increasing aridity gradient is not restricted to a few species, but is widespread among many different genera and families. No one species occurs along an entire aridity gradient in southwestern North America. More typically a species distribution will span a portion of the aridity transect. On the wetter and drier sites along such a transect, a species will often be replaced by other species within the same genus. Table 1 lists some of the more common genera whose species increase in leaf pubescence with aridity. In some cases there are only two species along the gradient and the genus extends over only a portion of the aridity transect, but in several instances different species of a genus may occur along the entire aridity gradient.

As the degree of leaf pubescence in these species increases, the percentage of light reflected by the leaf (leaf reflectance) also increases. This results in a decreased leaf absorptance. To illustrate this point, let us consider several species of *Salvia*. Along an increasing aridity gradient in southern California, *S. mellifera* occurs on the wettest sites, and is first replaced by *S. leucophylla*, and then by *S. apiana* on progressively drier sites (Munz, 1959). Figure 1 shows the leaf absorptance spectrum for these three *Salvia* species between 400 and 800 nm. *Salvia mellifera* with a glabrate, green leaf has an absorptance spectrum typical of most green leaves. On drier sites, it is replaced by the moderately pubescent, gray-leafed *S. leucophylla*. While the absorptance spectrum for *S. leucohpylla* is lower than *S. mellifera* over the visible (photosynthetically useful) wavelengths, the basic absorptance spectrum remains the same. In the heavily pubescent, white-leaved *S. apiana*, the percentage of light absorptance is much reduced below that of either *S. mellifera* or *S. leucophylla*.

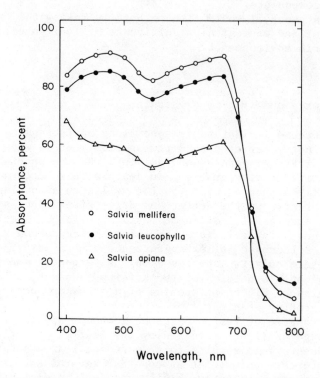

Figure 1. Monochromatic leaf absorptances between 400 and 800 nm of three *Salvia* species occurring along an aridity gradient.

Table 1. Genera which exhibit a variation in leaf pubescence along an aridity gradient in the Mohave and Sonoran Deserts.

genus	family
Arctostaphylos	Ericaceae
Artemisia	Asteraceae
Brickellia	Asteraceae
Ceanothus	Rhamnaceae
Condalia	Rhamnaceae
Encelia	Asteraceae
Enceliopsis	Asteraceae
Eriogonum	Polygonaceae
Franseria	Asteraceae
Kochia	Chenopodiaceae
Salvia	Lamiaceae
Sphaeralcea	Malvaceae
Tetradymia	Asteraceae
Viguiera	Asteraceae

Table 2. An elevational transect through Death Valley, California to demonstrate the replacement series in a group of closely related genera in the Asteraceae.

	life form	elevation (m)	absorptance (%)
Geraea canescens	annual	0-500	78
Encelia farinosa	shrub	300-800	29-81
Enceliopsis argophylla	herb	1000-1200	66
Encelia virginensis	shrub	900-1500	76

The trichomes responsible for a decreased leaf absorptance in *Salvia* are linear branched hairs. These hairs form a dense layer on the surface of the leaf, but because of the convoluted nature of the *Salvia* leaf the hairs appear as tufts.

The correlation between absorptance spectra and aridity in *Salvia* are typical of aridity gradients. Ehleringer (1980) has presented similar data for *Encelia* along a similar but more extensive aridity gradient in southern California. At the community level, Billings and Morris (1951) compared the spectral characteristics of species from several communities in the Great Basin Desert differing in aridity and they found that the average reflectance increased (absorptance decreased) as habitat aridity increased.

The pubescence which results in an increased reflectance or decreased absorptance is apparently a blanket reflector over the visible wavelengths (Ehleringer and Björkman, 1978). That is, the pubescence reflects all wavelengths between 400 and 700 nm equally well. This can be demonstrated in species of *Encelia* occurring along an aridity gradient in southern California. *Encelia farinosa* occurs in dry habitats and has pubescent leaves with a low leaf absorptance (Fig. 2). *Encelia californica* occurs at the moist end of the aridity gradient along the coast of southern California and has glabrate leaves with a spectrum typical of green leaves. When the hairs of *E. farinosa* leaves are removed and the absorptance spectrum of the now glabrous *E. farinosa* is remeasured, the spectrum is nearly identical to that of *E. californica* (Fig. 2). This suggests that in species replacements along aridity gradients (see Table 1), all species have leaves with the same basic absorptance pattern, but are covered with variable amounts of a blanket reflector.

Often, members of closely related genera occur parapatrically along aridity gradients. For example, *Encelia*, *Enceliopsis*, and *Geraea* are closely related genera occurring together in Death Valley. The lightly pubescent annual *Geraea canescens* is found at the lowest elevations (0-500 m) on the driest sites (Table 2). The leaves are green-gray with a leaf absorptance of 78% and it survives during the dry months as a seed. The heavily pubescent shrub, *Encelia farinosa*, occurs at higher elevations (300-800 m). The leaf absorptance of this species varies from 29 to 81% depending on aridity (a function of both elevation and season). *Enceliopsis argophylla*, a silver leaved perennial herb occurs on more moist sites above *E. farinosa* between 1000-1200 m. Leaf absorptance in this species is still higher at 66%. The shrub, *Encelia virginensis*, with lightly pubescent green leaves, has an absorptance of 76% and occurs at the highest, wettest elevations.

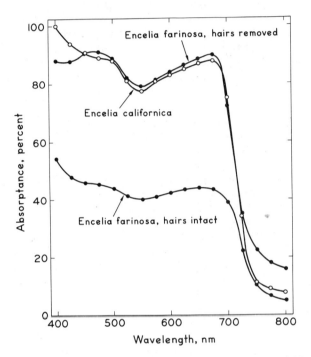

Figure 2. Monochromatic leaf absorptances between 400 and 800 nm for *Encelia farinosa* with hairs intact and with the hairs removed and the absorptance spectrum for an intact *E. californica* leaf. (From Ehleringer and Björkman 1978).

2.2 Leaf Absorptance and Life Form

The size, shape, and density of leaf hairs varies among arid land species. Both glandular and nonglandular leaf hairs occur in these species. Most of the nonglandular hairs found are simple and linear; however, the glandular hairs often range from simple to complex.

In annual species of arid lands, the hairs are most often simple and linear (Figs. 3-6). These hairs are of both glandular and nonglandular types and are always present in low densities relative to that found in other life forms. The hairs are large, often exceeding the thickness of the photosynthetic tissues by a factor of three to five times (Figs. 7-10).

Herbaceous perennial species may have leaf hairs similar to annuals or may have much more dense pubescence layers (Figs. 11-14). The dense hair layers are typically linear and nonglandular. Light microscope observations suggest that in leaves with thick indumentum, the hairs are dead and air filled. As the pubescence layer becomes thicker in these perennials, the epidermal surface becomes obscured (Figs. 12, 14). In thick pubescent leaves, the incoming light penetrates to the photosynthetic tissues only after multiple reflection.

Leaf pubescence in shrubs follows a pattern similar to that found in herbaceous perennials. However, there are probably more variations in leaf hair types in shrubs. For instance the linear hairs may be unbranched or branched as in *Salvia* (Figs. 15, 16) which also has less conspicuous glandular trichomes (Fig. 15).

Different hair types may vary in frequency as in *Encelia* which has two glandular and one nonglandular hair type (Figs. 17-20). In *Encelia* species with high leaf reflectances (e.g., *E. actonii, E. farinosa, E. palmeri*), most of the hairs are of the linear type. These linear hairs are greatly increased in length. In contrast, leaves of *E. frutescens* and *E. virginensis* have higher frequencies of glandular hairs, lower frequencies of linear hairs which remain relatively short, and low leaf reflectances. It is not possible to measure spectral characteristics of individual hairs, but the nonglandular hairs are probably responsible for the increased leaf reflectance.

The range of leaf absorptances for many species of Mohave and Sonoran Desert plants have been measured (Ehleringer, 1981a). When these data are arranged according to life form, only shrubs and herbs exhibit a wide range of leaf absorptances, even though all life forms may have pubescent leaves (Fig. 21). Tree and annual species all have high leaf absorptances. Small variations in these absorptances are attributable to changes in leaf thickness and thus to small changes in leaf transmittance. The wide range of leaf absorptances occurring in cacti results not from hairs, but rather from surface waxes and spines.

Additional types of trichomes other than hairs can also

ECOLOGY AND ECOPHYSIOLOGY OF LEAF PUBESCENCE IN DESERT PLANTS

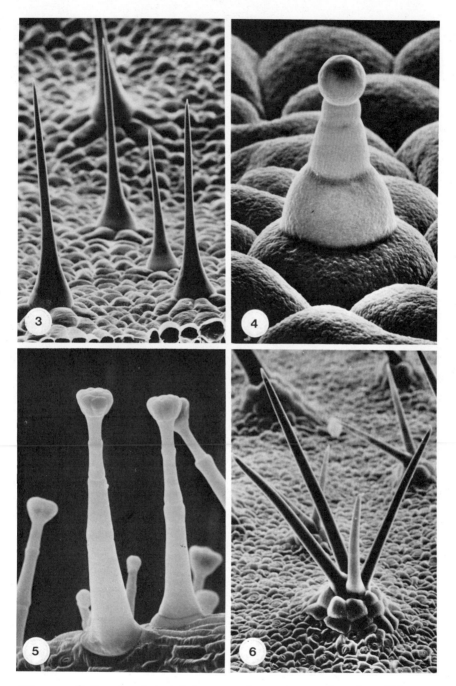

Figures 3-6. Leaf trichomes on desert annual. 3 - *Mohavea breviflora* (180X); 4 - *Abronia villosa* (740X); 5 - *Phacelia calthafolia* (240X); 6 - *Malvastrum rotundifolium* (200X).

Figures 7-10. Cross section of leaves of desert annuals. 7 - *Mohavea breviflora* (68X); 8 - *Abronia villosa* (54X); 9 - *Phacelia calthafolia* (76X); 10 - *Malvastrum rotundifolium* (60X).

ECOLOGY AND ECOPHYSIOLOGY OF LEAF PUBESCENCE IN DESERT PLANTS 123

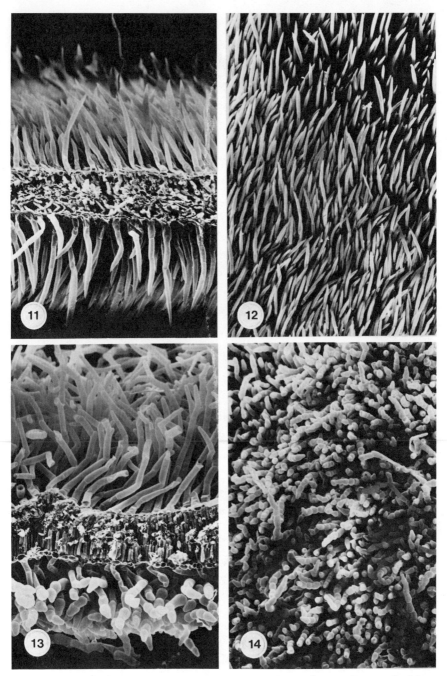

Figures 11-14. Leaf cross sections and surface views of desert perennial herbs. 11 - *Enceliopsis argophylla* (104X); 12 - *E. argophylla* (60X); 13 - *Psathyrotes ramosissima* (92X); 14 - *P. ramosissima* (64X).

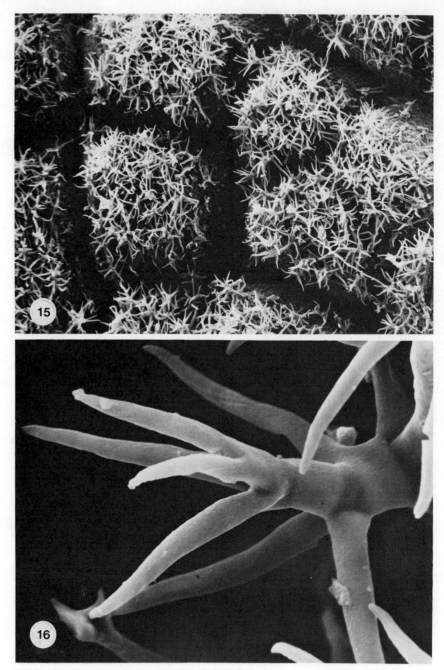

Figures 15-16. Leaf surface view and enlarged view of leaf trichomes in *Salvia leucophylla*. Figure 15 is 200X and Figure 16 is 2600X.

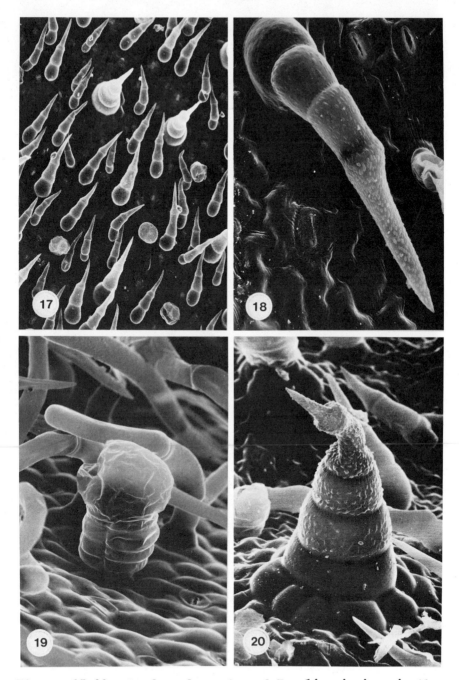

Figures 17-20. Leaf surface view of *Encelia virginensis* illustrating the density and diversity of trichomes and enlarged views of the predominant trichome types found in the species *Encelia*. Magnifications are 240X, 1080X, 880X and 650X, respectively.

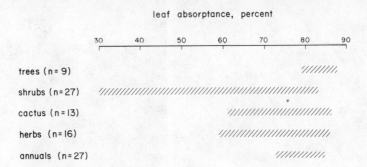

Figure 21. Ranges of leaf absorptances (400-700 nm) observed in a variety of plant species in the Mohave and Sonoran Deserts.

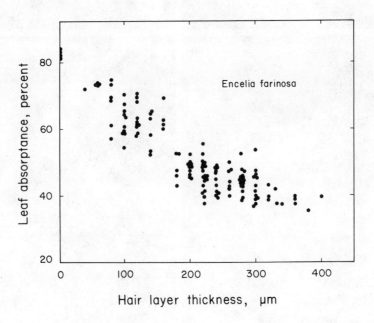

Figure 22. The relationship between increased leaf pubescence thickness and leaf absorptance (400-700 nm) in *Encelia farinosa* (From Ehleringer 1981b).

reduce leaf absorptance. Mooney et al. (1977) have shown that epidermal salt bladders in *Atriplex* may reduce leaf absorptance to about 40%.

The high leaf absorptances of tree and annual species may be the consequence of the water relations characteristic of those life forms. In the Mohave and Sonoran Desert, tree species are restricted to water courses with greater water availability. Annual species are typically ephemeral and complete their life cycle in mesic periods before the onset of extended drought.

2.3 Influence of Pubescence on Leaf Boundary Layer

Increases in the height of the pubescence layer should alter the boundary layer thickness, because the pubescence layer constitutes an additional component to the boundary layer. As the boundary layer becomes thicker, the rates of transfer of heat, water vapor, and CO_2 should decrease because they must diffuse across this layer.

Boundary layers are commonly quantified as resistances and not in terms of thickness. Resistance is a functional measure which indicates the impedance to the movement of substances across the layer. A typical leaf boundary layer resistance for water vapor is approximately 0.02 s mm^{-1} (Gates 1965). The total boundary layer resistance of a pubescent leaf will be

$$r_t = r_B + r_H \tag{1}$$

where r_T, r_B, and r_H are the total boundary layer resistance of a glabrous leaf, and additional boundary layer resistance due to the presence of hairs. Since heat and gases must diffuse across both components of the total boundary layer resistance, the impedances are in series and the total resistance is the sum of the components.

The additional boundary layer resistance from hairs can be calculated as

$$r_H = L/DA \tag{2}$$

where L is the height of the pubescence layer, A is the fractional area of the surface not occupied by hairs and available for diffusion, and D is the diffusion coefficient.

The thickness of the indumentum on leaves of varying pubescence has been measured in *Encelia farinosa* (Fig. 22). Leaf absorptance decreased asymptotically as the hair layer increased in thickness. Moderately pubescent leaves have hair layer thicknesses in the range of 100-200 µm. In extremely pubescent leaves with very low leaf absorptances, the pubescence may approach 400 µm. Assuming that A from equation 2 has a value in the neighborhood of one, the maximum increase in leaf boundary layer resistance for water vapor due to hairs will be approximately 0.017 s mm^{-1}. This represents almost a doubling of the

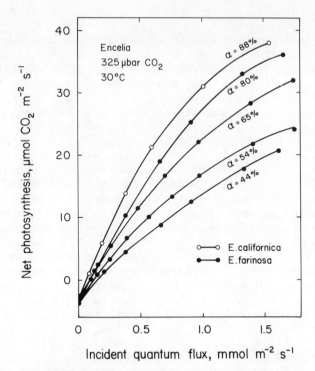

Figure 23. The relationships between quantum flux and photosynthesis in *Encelia* differing in leaf absorptance (α). (Based on data from Ehleringer and Mooney, 1978).

Table 3. Effects of changes in pubescence on leaf temperature and water loss in *Encelia farinosa* under mid-day summer conditions.

	green leaf	white leaf
absorptance (%)		
400-700 nm	85	40
400-3000 nm	50	17
leaf temperature (°C)	41.8	38.5
transpiration rate ($\mu g\ cm^{-2}\ s^{-1}$)	9.2	7.4

boundary layer resistance. However, functionally the effect on transpiration will be small.

The rate of water loss from a leaf will depend on the movement of water across both the boundary layer and stomatal pore resistances. A typical stomatal resistance for a pubescent *E. farinosa* leaf will be 0.5 s mm^{-1} (Ehleringer, 1977). This makes a total resistance to water loss of 0.52 s mm^{-1} if we neglect any additional resistances due to presence of the hairs. If the effects of hairs on boundary layer resistance are included, the total resistance becomes 0.537 s mm^{-1}; a change of only 3%. Consequently, the effects of hairs on boundary layer resistance are negligible in *Encelia* and probably also in all other arid land plant species.

2.4 *Effects of Pubescence on Ecophysiology*

A thick pubescence layer therefore has a significant effect on leaf spectral characteristics, but not on the total leaf layer resistances to water loss. Ehleringer and Mooney (1978) have addressed the question on how changes in leaf absorptance will affect leaf physiology.

With respect to transpiration and photosynthesis, both direct and indirect effects may occur. A decrease in solar radiation absorption will directly affect photosynthesis by reflecting quanta that might otherwise be used in carbon fixation. Leaf temperature is affected by decreasing the heat load on the leaf. Indirectly, a change in leaf temperature will affect both transpiration and photosynthesis. Transpiration will be modified since the saturation water vapor pressure at leaf temperature is a determining factor in the driving force for transpiration (Gates 1965). Photosynthesis will be modified as it is temperature dependent at both low and high light levels (Björkman 1973; Ehleringer and Björkman 1977, 1978).

In *Encelia farinosa* the effects of increased light reflectance by pubescence on photosynthesis are dramatic. In an analysis of photosynthesis light response curves with individual leaves of differing pubescence, Ehleringer and Mooney (1978) found that (1) *E. farinosa* leaves were not light saturated by noon irradiances and that (2) increased pubescence levels reduced photosynthetic rate at all light intensities (Fig. 23). The differences in rates of photosynthesis among the differentially pubescent leaves appeared to result solely from differences in leaf absorptance properties. When the rates of photosynthesis were expressed on an absorbed quantum flux basis rather than as incident quantum flux, the differences among the curves disappeared (Ehleringer and Mooney, 1978).

The effects of changes in leaf absorptance on leaf temperature and transpiration rate can be calculated using energy budget calculations (Gates, 1965; Ehleringer and Mooney, 1978). To illustrate this, let us consider the hypothetical situation of a "green" and "white" leaved *E. farinosa* (Table 3). The 400-700 nm leaf absorptances for "green" and "white" leaves are 85% and 40%, respectively. The total solar energy leaf absorptance (400-3000 nm) for these leaves would then be 50% and 17% (Ehleringer, 1981a). Under midday summer conditions (clear sky, 40°C air temperature, 20% relative humidity), the "green" leaf temperature is calculated to be 41.8°C and the "white" leaf 38.5°C. As a consequence, the transpiration rate of the "green" leaf is 9.2 µg cm^{-2} s^{-1} versus 7.4 g cm^{-2} s^{-1} for the "white" leaf. This 24% difference in water loss is likely to be of adaptive value to plants growing in arid habitats and may be one means of extending physiological activity into drought periods.

3. CONCLUSION

Leaf pubescence is abundant in desert plant species, and often increases in thickness in conjunction with increased aridity. A thick leaf hair layer results in reduced leaf absorptance (increased leaf reflectance). In the Mohave and Sonoran Deserts, these decreased leaf absorptances are found primarily among herbs and shrubs, but not among annuals or trees. Pubescence has little effect on the total leaf resistance to water loss, but does have a significant effect on leaf temperature, photosynthesis, and water loss through increased light reflectance.

4. REFERENCES

Billings, W.D., and R.J. Morris, 1951, Reflection of visible and infrared radiation from leaves of different ecological groups. Am. J. Bot. 38:327-331.

Björkman, O., 1973, Comparative studies on photosynthesis in higher plants, p. 1-63. *In:* Photophysiology, Vol. 8. Academic Press, New York.

Ehleringer, J., 1977, The Adaptive Significance of Leaf Hairs in a Desert Shrub. Ph.D. Thesis, Stanford University.

Ehleringer, J., 1980, Leaf morphology and reflectance in relation to water and temperature stress, p. 295-308. *In:* P. Kramer and N. Turner (eds.), Adaptations of Plants to Water and High Temperature Stress. Wiley-Interscience, New York.

Ehleringer, J., 1981a, Leaf absorptances of Mohave and Sonoran Desert plants. In Prep.

Ehleringer, J., 1981b, Boundary layer effects of pubescence in *Encelia farinosa*. In Prep.

Ehleringer, J. and O. Björkman, 1977, Quantum yields for CO_2 uptake in C_3 and C_4 plants: dependence on temperature, carbon dioxide and oxygen concentration. Plant Physiol. 59: 86-90.
Ehleringer, J. and O. Björkman, 1978, Pubescence and leaf spectral characteristics in a desert shrub, *Encelia farinosa*, Oecologia 36:151-162.
Ehleringer, J. and H.A. Mooney, 1978, Leaf hairs: effects on physiological activity and adaptive value to a desert shrub. Oecologia 37:183-200.
Ehleringer, J., O. Björkman and H.A. Mooney, 1976, Leaf pubescence: effects on absorptance and photosynthesis in a desert shrub. Science 192:376-377.
Gates, D.M., 1965, Energy, plants and ecology. Ecology 46:1-3.
Gausman, H.W. and R. Cardenas, 1969, Effect of leaf pubescence of *Gynura aurantiaca* on light reflectance. Bot. Gaz. 130: 158-162.
Gausman, H.W. and R. Cardenas, 1973, Light reflectance by leaflets of pubescent, normal, and glabrous soybean lines. Agron. J. 65:837-838.
Haberlandt, G., 1884, Physiologische Pflanzenanatomie. Engelmann, Leipzig.
Johnson, H.B., 1968, Pubescence as a Structural Feature of Vegetation. Ph.D. Thesis, Columbia University.
Johnson, H.B., 1975, Plant pubescence: an ecological perspective. Bot. Rev. 41:233-258.
Levin, D.A., 1973, The role of trichomes in plant defense. Quart. Rev. Biol. 48:3-15.
Mooney, H.A., J. Ehleringer and O. Björkman, 1977, The energy balance of leaves of the evergreen desert shrub *Atriplex hymenelytra*. Oecologia 29:301-310.
Munz, P.A., 1959, A California Flora. University of California Press, Berkeley.
Parkhurst, D.F., 1976, Effects of *Verbascum thapsus* leaf hairs on heat and mass transfer: a reassessment. New Phytol. 76: 453-456.
Pearman, G.I., 1966, The reflection of visible radiation from leaves of some western Australian species. Austr. J. Biol. Sci. 19:97-103.
Sayre, J.O., 1920, The relation of hairy leaf coverings to the resistance of leaves to transpiration. Ohio J. Sci. 20: 55-86.
Schimper, A.F.W., 1903, Plant Geography upon a Physiological Basis. Clarendon Press, Oxford.
Shreve, F. and I.L. Wiggins, 1964, Vegetation and Flora of the Sonoran Desert. Stanford University Press, Stanford.
Shull, C.A., 1929, A spectrophotometric study of reflection of light from leaf surfaces. Bot. Gaz. 87:583-607.

Sinclair, R. and D.A. Thomas, 1970, Optical properties of leaves of some species in arid south Australia. Austr. J. Bot. 18: 261-273.

Warming, E., 1909, Oecology of Plants: An Introduction to the Study of Plant Communities. Oxford University Press, London.

Wiegand, K.M., 1910, The relation of hairy and cutinized coverings to transpiration. Bot. Gaz. 49:430-444.

Wooley, J.T., 1964, Water relations of soybean leaf hairs. Agron. J. 56:569-571.

Wuenscher, J.E., 1970, The effect of leaf hairs of *Verbascum thapsus* on leaf energy exchange. New Phytol. 69:65-73.

BIOSYNTHESIS OF TERPENOIDS IN GLANDULAR TRICHOMES

Rodney Croteau and Mark A. Johnson

Institute of Biological Chemistry

and

Biochemistry/Biophysics Program
Washington State University
Pullman, Washington, 99164, U.S.A.

ABSTRACT

Monoterpenes (C_{10}), sesquiterpenes (C_{15}) and diterpenes (C_{20}) often constitute the major lipophilic substances associated with glandular trichomes, leading to a general tendency to regard these classical types of natural products as being closely related. While recent studies on the biosynthesis of mono-, sesqui- and diterpenes have revealed several common features in the molecular level construction of these compounds, there are distinct differences. Differences also exist with regard to sites of synthesis, turnover rates and other physiological aspects and with regard to probable function. In this review, the biochemistry of monoterpenes, sesquiterpenes and diterpenes is briefly described with particular reference to the similarities and differences in the origin of these classes of natural products.

1. INTRODUCTION*

The production of lipophilic materials by higher plants is often associated with the presence of glandular trichomes (Schnepf 1974; Fahn 1979). Most prominent among these lipophilic substances are the essential oils and resins, which consist primarily of terpenoid compounds. A precise distinction between the essential oils and the resins is difficult to make, and in the present context we will consider the essential oils to contain only volatile low molecular weight terpenoids [mostly monoterpenes (C_{10}) and sesquiterpenes (C_{15})] and the resins to consist of both volatile and non-volatile terpenoids [primarily diterpenes (C_{20})]. This review will thus be restricted to the origin of the C_{10}, C_{15}, and C_{20} classes of terpenes. Even within this limited objective, the early steps of terpene biosynthesis common to all organisms (Beytia and Porter 1976; Nes and McKean 1977; Qureshi and Porter 1980) and the biochemical polymerization of C_5 units (Poulter and Rilling 1978, 1980) will be excluded except for those specific aspects relevant to the synthesis of mono-, sesqui-, and diterpenes. General reviews on the biochemistry of plant terpenoids are available (Banthorpe and Charlwood 1980; Loomis and Croteau 1980), and detailed comprehensive treatments of the biochemistry of monoterpenes (Charlwood and Banthorpe 1978; Croteau 1980a), sesquiterpenes (Rucker 1973; Cordell 1976; Cane 1980b) and diterpenes (Hanson 1971; West 1980b) have recently been provided. Continuing coverage of the field is provided by the "Terpenoids and Steroids" and "Biosynthesis" Specialist Periodical Reports published by the Chemical Society. It is not our intent to provide comprehensive treatment here, but rather to concentrate on a limited series of more "classical" monoterpene, sesquiterpene and diterpene types, primarily of trichomal origin, and to focus specifically on the comparative aspects of the biochemistry of these compounds.

2. SITES OF SYNTHESIS

Numerous microscopic studies have indicated the presence of oils and resins in glandular trichomes, and analytical studies using sensitive gas chromatographic techniques have confirmed that glandular trichomes do, in fact, contain the monoterpenes,

* Abbreviations used throughout the text: MVA, mevalonic acid; IPP, isopentenyl pyrophosphate; DMAPP, dimethylallyl pyrophosphate; GPP, gernayl pyrophosphate; NPP, neryl pyrophosphate; LPP, linaloyl pyrophosphate; FPP, farnesyl pyrophosphate (all trans-isomer unless otherwise specified); GGPP, geranylgeranyl pyrophosphate.

sesquiterpenes and diterpenes characteristic of the plant
(Amelunxen and Arbeiter 1967; Sticher and Flück 1968; Amelunxen
et al. 1969; Malingre et al. 1969; Henderson et al. 1970; Dell
and McComb 1974, 1975, 1977; Heinrich 1977). While the terpenoid composition may vary both quantitatively and qualitatively
from one gland to the next, (Lemli 1957; Henderson et al. 1970)
and between different gland types of the same tissue (Amelunxen
et al. 1969; Heinrich 1977), there is little doubt that the
trichomes are the primary sites of terpene accumulation. There
has thus been a natural tendency to regard the trichomes as the
primary site of biosynthesis; however, attempts to demonstrate
this point conclusively have been hampered because of the
inability, in most cases, to achieve sufficiently high levels
of incorporation of exogenous labeled precursors into the
terpenes and because of the difficulties in obtaining sufficient
quantities of isolated intact trichomes for direct biosynthetic
experimentation.

In spite of these experimental limitations, evidence has
indicated that the trichomes are capable of incorporating
exogenous precursors such as acetate and MVA into the constituent
terpenoids (Michie and Reid 1959; Reid 1979; Loomis and
Croteau 1973; Verzár-Petri and Then 1975; Dell and McComb
1978), demonstrating that the requisite biosynthetic capability
is present in such structures. More recently, it was demonstrated
that the isolated epidermis from *Majorana hortensis* leaf (containing the intact trichomes) possessed the same ability to
convert [U-^{14}C] sucrose to constituent monoterpenes as did the
whole leaf, while the isolated mesophyll layer was inactive
in synthesizing monoterpenes from the same precursor (Croteau
1977). Thus, it would appear that the epidermis, if not the
glandular trichomes themselves, does represent the primary site
of terpene synthesis.

The generally low levels of incorporation of exogenous
precursors into the mono-, sesqui- and diterpenes relative to
other terpene classes (e.g., triterpenes) suggests that the sites
of synthesis of the lower terpene classes are compartmentalized
and thus not as assessible to exogenous precursors as are more
"primary" biosynthetic processes (Nicholas 1962a, b; 1964;
Ruddat et al. 1965; Loomis and Croteau 1973; Dell and McComb
1978). Compartmentation of terpenoid metabolism in plants is
well recognized as a feature of regulation (Rogers et al. 1968;
Goodwin 1977; Rappaport and Adams 1978; West et al. 1979) and
the morphology of most types of glandular trichomes does suggest
a degree of isolation from the rest of the plant. Terpenoids
that are formed in the glandular epidermis of flowers, as opposed
to trichomes, incorporate exogenous precursors rather efficiently
(Francis 1971), as do certain terpenoid classes (such as
iridoid monoterpenes) that are not sequestered in glandular
structures (Banthorpe et al. 1972a; Inouye 1978), thus sup-

porting the concept of the trichomes as the physical basis of compartmentation for the biosynthesis of the types of terpenes described here. The situation is further complicated, however, by observations suggesting that degrees of compartmentation exist within the trichome itself. Thus, incorporation of radioactive glucose, sucrose or CO_2 into the terpenes of peppermint results in far greater labeling of monoterpenes than sesquiterpenes (the ratio is about 20:1, reflecting the approximate natural proportion of mono- and sesquiterpenes in the plant) (Croteau et al. 1972a,b). Conversely, exogenous MVA is a far more efficient precursor of sesquiterpenes than of monoterpenes in this tissue (i.e., about 0.33% incorporation versus approximately 0.03% incorporation into monoterpenes) (Croteau and Loomis 1972). Differential labeling between monoterpenes and sesquiterpenes has also been observed in other plants (Regnier et al. 1968; Loomis and Croteau 1973; Banthorpe and Ekundayo 1976; Gleizes 1978; Bernard-Dagan et al. 1979). Thus, the evidence would appear to suggest distinct sites for monoterpene and sesquiterpene biosynthesis differing in their accessibility to exogenous precursors. The limited data available suggests that the sites of diterpene biosynthesis may also differ in accessibility from those of mono- and sesquiterpene biosynthesis (Nicholas 1964). Thus, while the trichomes may be regarded as the primary site of terpene biosynthesis and the principal physical basis for certain compartmentation phenomena, it is also clear that multiple compartmentalized biosynthetic sites probably exist within the "secretory cells" of the trichomes.

Too little information is presently available to permit correlation of postulated compartments with structural entities within the oil glands (Loomis and Croteau 1973; Heinrich 1979), and attempts to relate cytological observations with their physiological and biochemical consequences are largely conjectural. It does seem likely, however, that the apparent degree of isolation of the biosynthetic apparatus within structures that generally lack photosynthetic capability and, perhaps, an adequate supply of oxygen (Burmeister and von Guttenberg 1960) would make terpene biosynthesis highly dependent on nearby cells for both energy and structural precursors (Loomis and Croteau 1973). Experimental attempts to approach this question through *in vivo* co-feeding of MVA along with fermentable substrates have supported this concept (Croteau et al. 1972b).

Tracer studies have indicated that immature leaves, presumably bearing "immature" glandular trichomes, are far more efficient at terpenoid synthesis from exogenous precursors than are mature, fully expanded leaves (Battaile and Loomis 1961; Croteau 1977; Dell and McComb 1978; Croteau et al. 1980c). It is equally clear from analytical studies that, at least under certain circumstances, mature leaves continue to accumulate terpenes, suggesting that glands may be capable of synthesizing terpenes from stored substrates once access to exogenous substrate is restricted (Burbott and Loomis 1967). The probable consequences of compartmentation of terpene biosynthesis within the glandular trichomes have been discussed previously (Banthorpe et al. 1972a; Loomis and Croteau 1973; Charlwood and Banthorpe 1978), yet many of the questions raised in these earlier papers concerning the relationship of oil glands to terpene biosynthesis have remained unanswered.

3. TERPENE STRUCTURES AND THE ISOPRENE RULE

Several hundred naturally occurring monoterpenes are presently known (Devon and Scott 1972) and they can be divided into several structural types of which the cyclohexanoid family contains, by far, the greater number of individual compounds. A few representative examples are illustrated in Figure 1. The cyclohexanoid monoterpenes are subdivided into monocyclic and bicyclic types, and within these categories further divisions are made on the basis of skeletal arrangements (p-menthanes, pinanes, bornanes, etc.). The chemistry of monoterpenes has been reviewed most recently by Whittaker (1972).

The sesquiterpenes comprise some 200 skeletal types (Andersen et al. 1978; Cane 1980b), and the several thousand individual sesquiterpenes (Devon and Scott 1972) represent one of the largest and most diverse families of natural products. A few representative sesquiterpenes are illustrated in Figure 2, and extensive surveys of the structures and chemistry of the sesquiterpenes are available in the literature (Herout 1971; Roberts 1972). The diversity of carbon skeletons among the sesquiterpenes appears to greatly exceed that of the mono- and diterpenes, although it is clear that in the mono- and diterpene series far more theoretical structural possibilities exist than have actually been found so far in nature [see, for example, Smith and Carhart (1976)]. At least a thousand diterpenoid compounds have now been identified (Devon and Scott 1972) of which only a few examples are illustrated here (Fig. 3). Several aspects of diterpene chemistry have been reviewed by Hanson (1972), Nakanishi (1974), and White and Manchard (1978).

A catalog of many terpene structures and schemes for their biogenetic relationships have been provided by Devon and Scott (1972), and the taxonomic distribution of mono-, sesqui- and diterpenes among higher plants has been described by Tetenyi (1970), Darnley-Gibbs (1974) and Banthorpe and Charlwood (1980).

The terpenes of higher plants are most often cyclic, contain relatively few simple functional groups (hydroxyls, carbonyls, double bonds, etc.), and largely represent variations on a limited number of skeletal themes. As would be expected, the mono-, sesqui- and diterpenes exhibit a number of similar physical

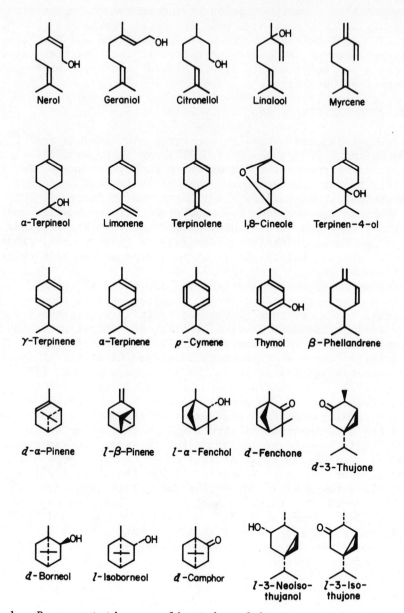

Fig. 1. Representative acyclic and cyclohexanoid monoterpenes.

BIOSYNTHESIS OF TERPENOIDS IN GLANDULAR TRICHOMES

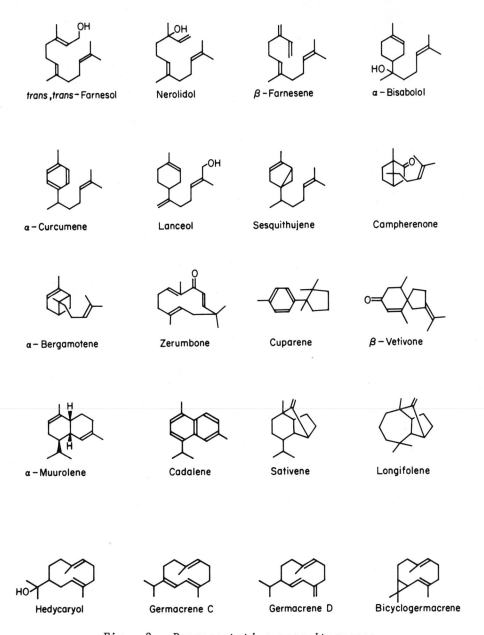

Fig. 2. Representative sesquiterpenes.

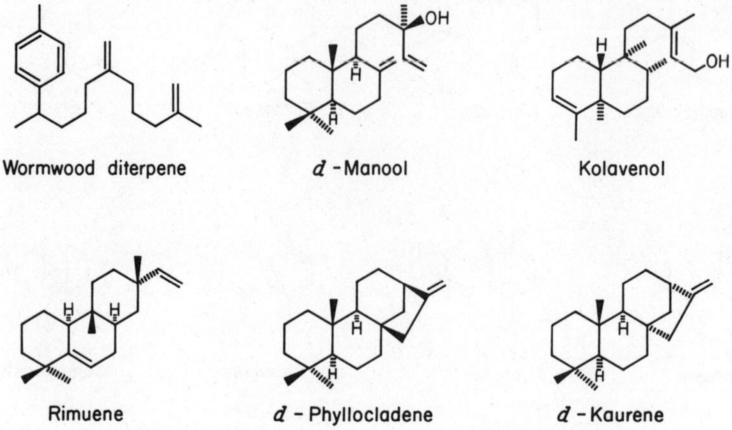

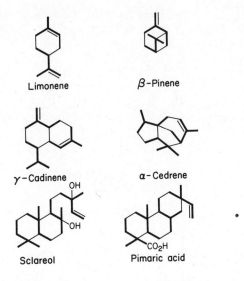

Fig. 3. Representative diterpenes.

Fig. 4. Representative mono-, sesqui- and diterpenes and their division into isoprene units in a head-to-tail pattern.

and chemical properties. The boiling point range increases accordingly with size, accounting for the presence of mono- and sesquiterpenes in the essential oils, and diterpenes in the non-volatile fraction of resins. Nonvolatile conjugates (glycosides, esters) and highly oxygenated derivatives of monoterpenes and sesquiterpenes do occur [e.g., iridoid monoterpenes (Bobbitt and Segebarth 1969) and sesquiterpenes lactones (Yoshioka et al. 1973)], but these and other specialized classes of the higher plant terpenes, such as gibberellins, will be mentioned here only in passing.

While at first approach, the mono-, sesqui- and diterpenes may appear to present a bewildering array of acyclic and, more commonly, cyclic systems comprising a variety of derivatives and isomers, it was clear to early workers (Wallach 1914) that most terpenes could be hypothetically constructed by sesquential condensation of isoprene (isopentane) units in a head-to-tail pattern (Fig. 4), thus providing some degree of classification. Many terpenes were subsequently shown not to obey this simple generalization, yet all could be rationalized by the "biogenetic isoprene rule" in which Ruzicka (Ruzicka et al. 1953; Ruzicka 1959, 1963) proposed that naturally occurring terpenes are derived directly, or indirectly (i.e., via rearrangement), from common acyclic C_{10}, C_{15} and C_{20} isoprenoid precursors. This hypothesis, which ignores the precise character of the acyclic biological precursors (and intermediates), is based largely on cationic mechanisms for construction of the various skeletal types via a series of internal additions (i.e., electrophillic attack of a positive center on a double bond) hydride shifts and rearrangements. Once the parent cations of the various structural types have been thus generated, the ionic reactions would be terminated by the addition of a nucleophile (e.g., OH^- from water) to give the corresponding alcohol, or where possible, by elimination of a proton to give the corresponding olefin. Subsequent secondary transformations of these parent compounds could then give rise to the great variety of terpenes encountered in plants. The biogenetic isoprene rule and its later extensions and elaborations thus provides a useful, and unifying, classification of the terpenes based on biogenetic considerations. This hypothesis has guided both structural investigations and studies on biosynthetic pathways and mechanisms, as will be elaborated in the following sections.

4. BIOSYNTHESIS

4.1 *Terpene Precursors*

In spite of the generally low levels of incorporation of exogenous precursors into the monoterpenes of higher plants, the mevalonoid origin of these compounds has been clearly demonstrated

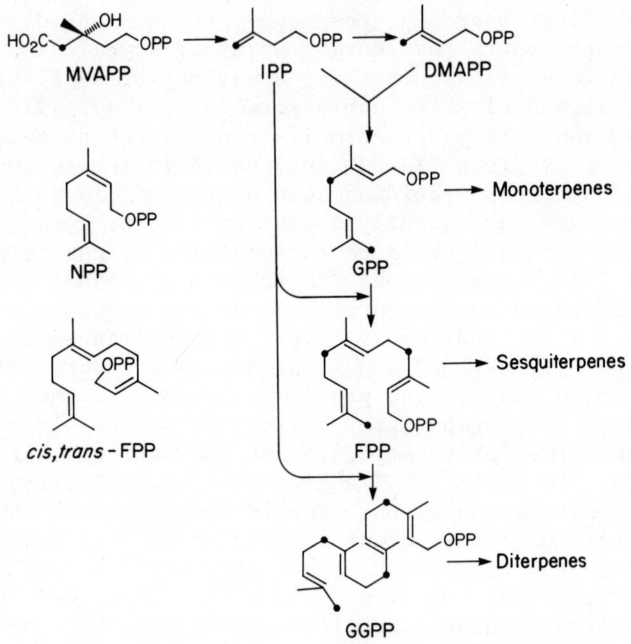

Fig. 5. Outline of the early steps of the isoprenoid pathway leading from mevalonic acid pyrophosphate to acyclic precursors. Closed circles indicate the position of labeling from [2-^{14}C]mevalonic acid.

by *in vivo* labeling experiments (Banthorpe et al. 1972a). In the case of sesquiterpene and diterpene biogenesis, fungal systems have proven to be far more suitable for the purpose of delineating the mevalonate pathway because of higher incorporation levels, yet the origin of a sufficient number of higher plant sesquiterpenes and diterpenes (Souchek 1962; Ruddat et al. 1965; Breccia and Badiello 1967; Zabkiewicz et al. 1969; Biollaz and Arigoni 1969; Corbella et al. 1969; Croteau and Loomis 1972) has been examined *in vivo* such that the matter is no longer in doubt. A curious result of exogenous tracer studies using basic terpenoid precursors is the commonly observed asymmetric labelling of the C_5 units of the derived terpene (i.e., the IPP derived portion of the molecule is generally more heavily labelled with tracer than the portion of the molecule derived from DMAPP) (Banthorpe et al. 1972a; Loomis and Croteau 1973; Charlwood and Banthorpe 1978). Such labelling patterns have been observed even with $^{14}CO_2$ as the precursor (Croteau et al. 1972a, Wuu and Baisted 1973; Banthorpe et al. 1975). Asymmetric labelling has been the object of considerable discussion and is generally attributed to the participation in the reaction of an endogenous pool of DMAPP (Loomis and Croteau 1973; Banthorpe and Charlwood 1980). As the equilibrium of the IPP to DMAPP isomerization (Fig. 5) lies to the right (Shah et al. 1965) and this step is unlikely to be rate limiting, it seems probable that the phenomenon is ubiquitous in isoprenoid biosynthesis but is only discernable when the incorporation of labelled precursor is extremely low, as is often the case with the terpenes under consideration here.

While the biogenetic isoprene rule was originally formulated with allylic C_{10}, C_{15} and C_{20} prenols as precursors of the mono-, sesqui- and diterpenes respectively, the corresponding pyrophosphate esters, GPP, FPP and GGPP (Fig. 5) are now considered to be the more appropriate precursors of terpene compounds. For obvious steric reasons, GPP cannot cyclize directly to cyclohexanoid monoterpenes, and so the *cis*-isomer of GPP (NPP, Fig. 5) has traditionally been regarded as the more likely direct acyclic precursor of these cyclic compounds (Valenzuela et al., 1966; Loomis, 1967). Similarly, *cis, trans*-FPP (Fig. 5) has been proposed as the likely precursor of certain sesquiterpene classes that cannot be formed directly from the *trans, trans*-isomer (Parker et al. 1967; Rücker 1973). Isomers of GGPP have only rarely been suggested as precursors of diterpenes (Oehlschlager and Ourisson 1967), and the all *trans* compound is now considered to be the general acyclic precursor of the diterpene series. However, in the case of monoterpene and sesquiterpene biosynthesis there has been considerable speculation as to how the proposed *cis*-isomers might arise, including direct *cis*-condensation of IPP (Jedlicki et al. 1972; Perez et al. 1980) and isomerization schemes involving aldehydes as intermediates [i.e., GPP→geranoil→geranial→neral nerol→NPP (Charlwood and Banthorpe 1978)] (For critical discussion of these schemes see Cane 1980a). Recently, studies with cell-free preparations have shown that GPP is transformed to cyclic monoterpenes without preliminary conversion to NPP (Croteau and Karp 1979a; Croteau and Felton 1980b; Croteau 1980a; see also following section). These studies indicate that the monoterpene cyclase enzymes are capable of transforming GPP to an intermediate (probably enzyme-bound) that is sterically capable of cyclization, removing the need to invoke free NPP as the direct precursor of cyclic monoterpenes. A similar situation may apply in the biosynthesis of cyclic sesquiterpenes from *trans, trans*-FPP, eliminating the proposed intermediacy of the *cis, trans*-isomer (see following section). Thus, it now seems likely that the mono-, sesqui- and diterpenes are all formed, more or less directly, from the normal C_{10}, C_{15} and C_{20} *trans*-allylic pyrophosphates that arise from the common C_5 elongation process catalyzed by prenyl transferase (Fig. 5). The prenyl transferase reaction (Poulter and Rilling 1978, 1980) and the early steps of the isoprenoid pathway common to all terpene types (Beytia and Porter 1976; Qureshi and Porter 1980) have been described in detail elsewhere, and are illustrated in outline form in Figure 5. A detailed discussion of the stereochemical aspects of prenyl pyrophosphate metabolism has recently been provided by Cane (1980a).

While it is not possible to discuss the general isoprenoid biosynthetic pathway in any detail here, it is important to point out that the monoterpenes, sesquiterpenes and diterpenes diverge at important junctures of this common pathway. Thus, the C_{10} intermediate GPP can divert to the monoterpenes or undergo elongation to the C_{15} intermediate. FPP can give rise to the sesquiterpenes, undergo dimerization to squalene and thence yield the triterpenes, or be further elongated. GGPP can give rise to the diterpenes, dimerize to the tetraterpenes (e.g., carotenoids) or undergo further elongation to polyprenols. Product formation at these various branch points might thus be expected to be under rather strict metabolic regulation. However, it seems quite probable that the entire terpenoid pathway is not operative in all terpene biosynthetic sites and that control could be to a significant degree exerted by compartmentation of chain-length specific prenyl transferases with the appropriate C_{10}, C_{15} and C_{20} cyclases and modifying enzymes. Within a compartment, however, strict "branch point regulation" of the type implied above, could be important. Regulation of metabolism in the diterpene series has been discussed (West et al. 1979), and the prenyl transferases considered to be involved is the synthesis of diterpenes and gibberellins are rather specific for the construction of GGPP (West 1980b). Most prenyl transferases that have been studied preferentially synthesize FPP (Poulter and Rilling 1980), primarily as the precursor of squalene but also suitable for the synthesis of sesquiterpenes. Tissues specialized for the synthesis of monoterpenes might therefore be expected to possess a prenyl transferase specific for the condensation of DMAPP and IPP to GPP. Such C_{10} specific prenyl transferases have not yet been demonstrated in higher plants, but an enzyme of this specificity was recently isolated from a bacterium (Sagami et al. 1978).

4.2 *Cyclization Processes*

Most monoterpenes, sesquiterpenes, and diterpenes, are cyclic compounds, and the biosynthesis of these terpene classes can be conveniently divided into two separate processes; the cyclization of the various acyclic precursors, GPP, FPP and GGPP, to the appropriate skeletal type, and the secondary metabolic transformations of the parent cyclic compound. The various cyclization processes have received the greatest amount of experimental attention, because it is the cyclization step which determines the basic structural character of the terpenes a given organism is capable of producing. In almost all instances terpene cyclization processes can be rationalized in terms of carbonium ion mechanisms [as put forward in the biogenetic isoprene rule (Ruzicka et al. 1953)], with the cyclization being initiated either by loss of the terminal pyrophosphate or by protonation of a double of the acyclic precursor, and the outcome of the reaction being guided by stereoelectronic factors. The commonly accepted role of the cyclase enzymes is thus to exploit the inherent reactivity of the allylic pyrophosphate substrate, and, through various types of interaction with the substrate, dictate the regio- and stereochemistry of the product.

In the context of cyclization processes, the stereochemistry of the terpenes deserves some comment. The homogenous absolute stereochemistry observed in the triterpenes is not observed in the lower terpenes classes. Thus, enantiomeric skeletal types (p-menthanes, pinanes, bornanes, fenchanes) are common among the monoterpenes of higher plants (Karrer 1958; Plouvier 1966; Devon and Scott 1972) and they occasionally co-occur in the same species. The classical example of the latter is a-pinene which is seldom found in optically pure form (Banthorpe and Whittaker 1966; Zavarin 1970). Conversely, β-pinene is generally found as the optically pure l-isomer. Enantiomeric skeletal modifications are also naturally occurring among the sesquiterpenes (d- and l-a-bisabolol) and diterpenes (d- and l-kaurene) of higher plants, although in these classes the occurrence of enantiomeric forms is, apparently, less common (Oehlschlager and Ourisson 1967; Andersen et al. 1978). Why such a variety of antipodal forms should exist among the monoterpenes is not immediately apparent. It seems likely that an initial (and single) binding of the enzyme to GPP could, to a large degree, determine the stereochemical outcome of monoterpene cyclizations (Croteau and

Karp 1979a; Croteau et al. 1980b) and a similar argument has
been put forward for the construction of certain sesquiterpenes
(Arigoni 1975; Andersen et al. 1978) and diterpenes (Eschenmoser
et al. 1959). However, for more complex sesqui- and diterpene
cyclizations involving stereochemical outcome may be required.
Thus, it might be argued that enantiomers are most likely to
be encountered among those terpene types whose skeletal construction requires minimum intervention of the enzyme.

    a. *Monoterpenes*. Ruzicka's general scheme for the biogenesis of cyclic monoterpenes from a common acyclic presursor is shown in Figure 6. While Ruzicka was unable to formulate the C_{10} prenyl pyrophosphate as the active biochemical precursor in 1953 (Ruzicka et al. 1953), he did appreciate the steric impossibility of direct cyclization of a precursor with the *trans* configuration (i.e., a geranyl derivative) and therefore, ascribed a central role to the *cis* (neryl) isomer as the most appropriate acyclic progenitor of cyclohexanoid monoterpenes. Later workers (Valenzuela et al. 1966; Loomis 1967) were to suggest the corresponding pyrophosphate, NPP, as the most probable direct biosynthetic precursor of cyclic monoterpenes. This concept has since gained wide acceptance, although the evidence for NPP as an obligate precursor is not compelling and reaction schemes for the cyclization of GPP via enzyme-bound intermediates (and without the involvement of any other prenyl pyrophosphate) can be readily envisioned.

    In the hypothetical context of monoterpene cyclizations, the origin of NPP has assumed considerable importance and several possible biosynthetic routes to NPP have been proposed [see Cane (1980a) for discussion]. In addition to NPP, linaloyl pyrophosphate (LPP) has also been proposed as the immediate acyclic precursor of cyclohexanoid monoterpenes (Attaway et al. 1966; Suga et al. 1980) (for structure of linaloyl see Fig. 1) and in this instance also there is no stereochemical obstruction to direct cyclization. Schemes for the biosynthesis of LPP have also been put forward (Attaway et al. 1967), but conclusive experimental evidence for such hypotheses, and for the obligate role of LPP in cyclization processes, is lacking. Of greater significance than the hypothetical origins of NPP or LPP are recent studies with several partially purified monoterpene synthetases which demonstrate that GPP is the preferred substrate for cyclization, in the absence of any competing reactions and any detectable interconversion of GPP, NPP and LPP (Croteau and Karp 1979a; Croteau et al. 1980b). Even in relatively crude cell-free preparations in which competing activities (e.g., phosphohydrolases) obscure substrate specificity studies, and where NPP and LPP are apparently more efficient precursors than GPP (Croteau and Karp 1977a), the fact that GPP is cyclized (in the absence of detectable interconversion) strongly suggests

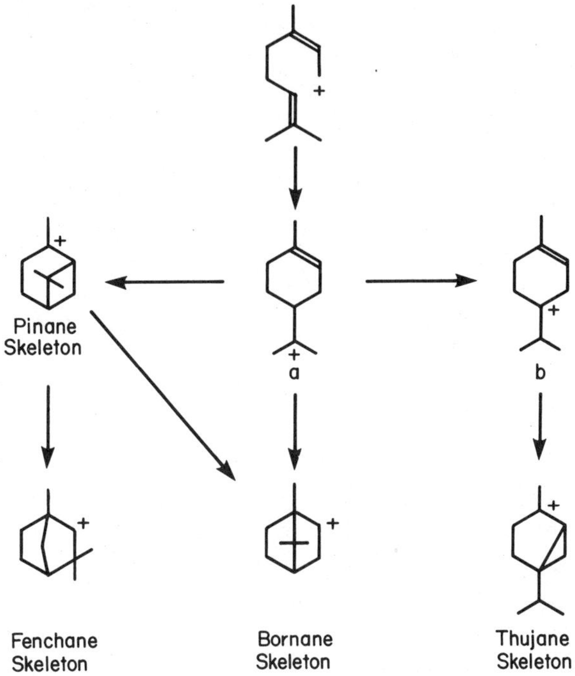

BIOSYNTHESIS OF TERPENOIDS IN GLANDULAR TRICHOMES

Fig. 6. Postulated ionic mechanism for the formation of monoterpene skeleta via the α-terpinyl cation (a) and the terpinen-4-yl cation (b) (after Ruzicka et al. 1953).

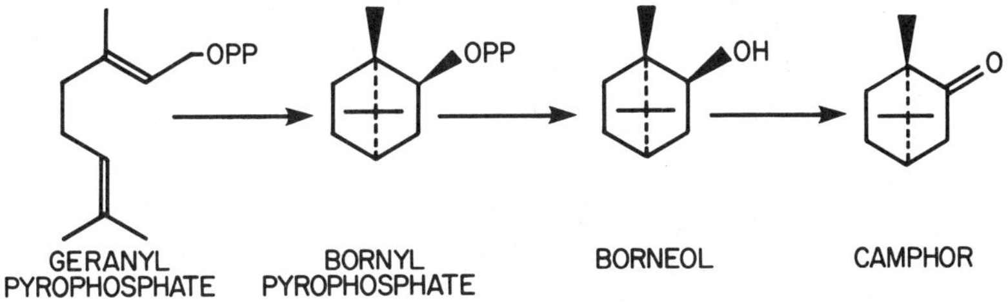

Fig. 7. Pathway for the enzymatic conversion of geranyl pyrophosphate to bornyl pyrophosphate, borneol and camphor. Structures illustrated are the d-isomers.

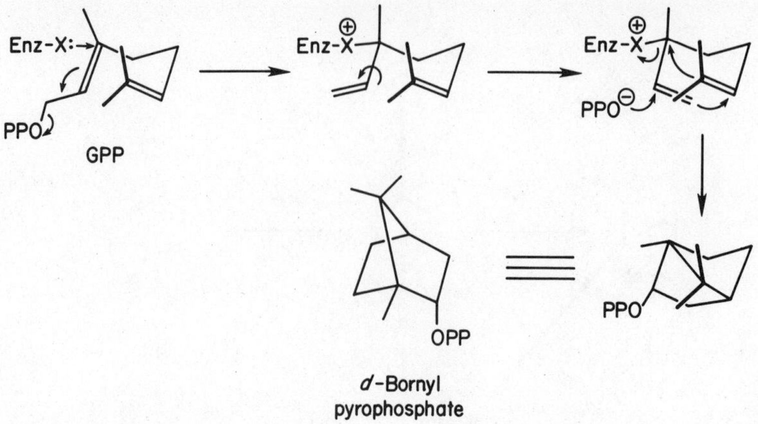

Fig. 8. Possible mechanism for the cyclization of geranyl pyrophosphate to d-bornyl pyrophosphate via an enzyme-bound linaloyl intermediate.

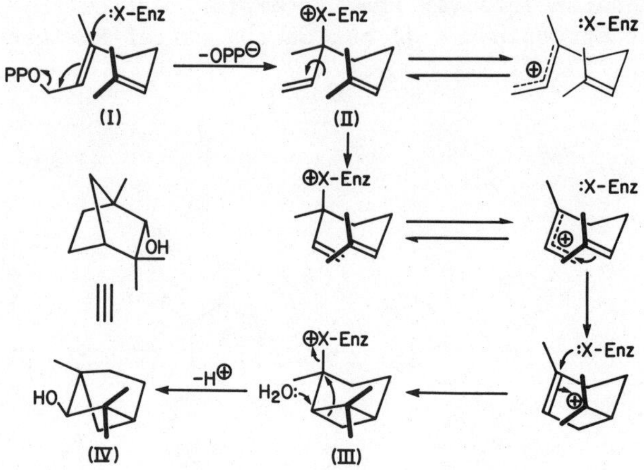

Fig. 9. Possible mechanism for the cyclization of geranyl pyrophosphate (I) to l-endo-fenchol (IV) via enzyme-bound linaloyl (II) and pinyl (III) intermediates.

that the biosynthesis of cyclic monoterpenes can be sustained *in vivo* from GPP without invoking other obligate free intermediates.

Perhaps the best evidence for the direct role of GPP in monoterpene cyclizations comes from studies on the biosynthesis camphor by cell-free preparations from sage *(Salvia officinalis)* (Croteau and Karp 1976b, 1979a). This multistep pathway (Fig. 7) involves the initial cyclization of the acyclic precursor to the novel intermediate d-bornyl pyrophosphate followed by hydrolysis to d-borneol and dehydrogenation to d-camphor. l-Camphor is formed by a similar pathway with l-bornyl pyrophosphate as an intermediate (Croteau and Karp 1977b). NPP was the more efficient precursor of bornyl pyrophosphate in crude cell-free preparations, but as purification of the bornyl pyrophosphate synthetase proceeded, an increase in substrate preference for GPP was observed under conditions where no interconversion of GPP and NPP occurred (Croteau and Karp 1979a). This apparent change in substrate preference was subsequently shown to result from the presence of phosphohydrolases in the crude extract which rapidly hydrolyzed GPP to the inactive monophosphate (Croteau and Karp 1979b). Only on the removal of the contaminating phosphohydrolases was the true substrate preference for GPP over NPP apparent. The partially purified bornyl pyrophosphate synthetase from sage had an apparent molecular weight of 95,000, and required Mg^{++} for catalytic activity. The enzyme exhibited a pH optimum at 6.2 and was strongly inhibited by thiol directed reagents such as p-hydroxymercuribenzoate and by the active serine directed reagent diisopropylfluorophosphate. Bornyl pyrophosphate synthetase was the first monoterpene synthetase to be isolated free from competing phosphohydrolases, and the first to show a preference for GPP as substrate. This finding necessitated reconsideration of the role of the cyclase enzymes in the biosynthesis of cyclic monoterpenes, as it implied a multistep reaction sequence in which the *trans*-isomer must initially be transformed to an intermediate stereochemically suitable for cyclization. To carry out such a transformation, it was suggested that the substrate first react with the enzyme, with concomitant loss of pyrophosphate, to afford a bound linaloyl intermediate which permits rotation around the C(2) - C(3) bond (Croteau and Karp 1979a) (Fig. 8). Once the appropriate conformation is achieved, cyclization and release of the stereochemically appropriate product could occur with insertion of the pyrophosphate and removal of the enzyme as indicated.

The biosynthesis of the rearranged bicyclic monoterpene fenchol was recently demonstrated in cell-free preparations from Fennel *(Foeniculum vulgare)* (Croteau et al. 1980a). In this instance the free alcohol l-*endo*-fenchol appears to be the ultimate product of the enzyme, rather than the pyrophosphate, indicating a different mode of terminating the cyclization. In this case, the partially purified fenchol synthetase was also shown to prefer GPP over NPP as substrate, but the specificity for the *trans*-isomer was not as pronounced as that observed for bornyl pyrophosphate synthetase (Croteau et al. 1980b). A possible mechanism for the conversion of GPP to *l-endo*-fenchol via an enzyme-bound linaloyl cation was proposed (Croteau et al. 1980b) (Fig. 9). This scheme involves the rearrangement of a pinyl intermediate (Fig. 9, III), consistent with the labelling pattern of the product obtained from specifically labelled precursors (Croteau et al. 1980a), and it accounts for the correct stereochemistry while requiring only a single nucleophile and a single point of attachment. The fenchol synthetase exhibits a molecular weight of about 60,000, a pH optimum near 7.0, and it requires Mn^{++} for catalytic activity.

The biosynthesis of the bicyclic hydrocarbons, α- and β-pinene (Fig. 1), has also been demonstrated in cell-free preparations (Croteau and Karp 1976a; Chayet et al. 1977). Like other monoterpene synthetases, the pinene synthetase activities require a divalent cation, exhibit a pH optimum between 6 and 7, and are inhibited by thiol-directed reagents. In no case has a pinene synthetase been purified to a sufficiently high degree to allow studies of the enzyme's properties or mechanism of catalysis. A mechanism similar to that illustrated in Figure 9 could, however, readily account for the formation of α- and β-pinene from intermediate III (Croteau et al. 1980b). Two distinct α-pinene synthetases have been demonstrated in extracts of sage leaves (Gambliel and Croteau unpublished). While the stereochemistry of the conversion of the substrate to the appropriate enzymebound linaloyl intermediate, followed by cyclization and double bond formation with the accompanying hydride shift. Similar enzyme stablized intermediates have been postulated by Cori and associates (Chayet et al. 1977; Bunton and Cori 1978) to rationalize the conversion of both GPP and NPP to monoterpene hydrocarbons by enzyme preparations from *Citrus*. The means by which GPP is cyclized on the molecular level, and the nature of the biosynthetic intermediates (represented here by enzyme-bound carbonium ions), is not clear. Reactive intermediates closely associated with, or bound to, the enzyme would appear necessary in order to promote cyclization and confer the regio- and stereospecificities observed. To approach such fundamental questions will require more detailed studies of the cyclase enzymes than have, heretofore, been carried out.

    b. *Sesquiterpenes*. As in the case with cyclohexanoid monoterpenes, several types of cyclic sesquiterpenes have traditionally been regarded on stereochemical grounds as being derived from a *cis*-allylic pyrophosphate precursor (i.e., *cis*-Δ^2, *trans*-Δ^6 farnesyl pyrophosphate rather than the common all *trans*-isomer). The finding cyclic monoterpenes traditionally regarded as being derived from NPP are, in fact, derived from

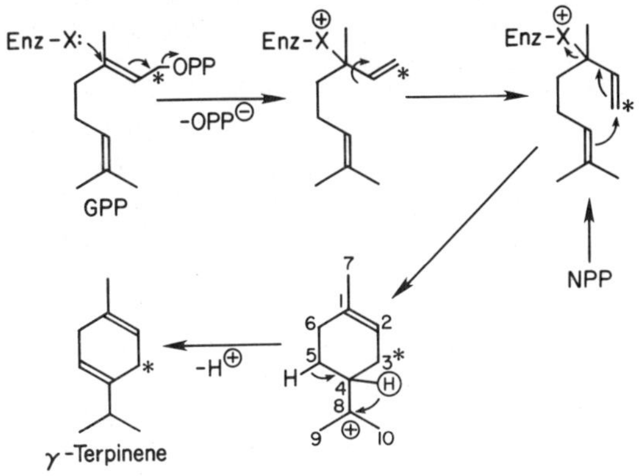

Fig. 10. Possible mechanism for the cyclization of geranyl pyrophosphate or neryl pyrophosphate to γ-terpinene. Asterisks indicate the position of tritium from the [1-³H]-labeled acyclic precursors.

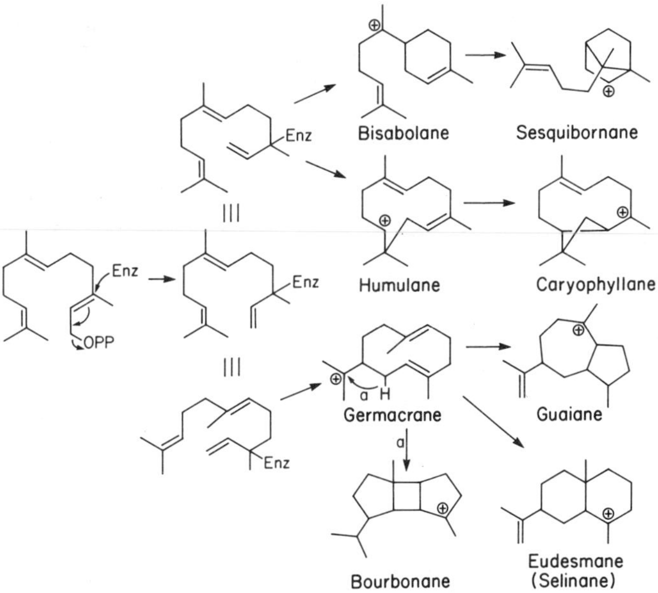

Fig. 11. Hypothetical formation of representative sesquiterpene skeleta by cyclization of farnesyl pyrophosphate through a nerolidyl intermediate. Stereochemical considerations are ignored.

the ubiquitous GPP raises an obvious question about an obligate role of *cis, trans*-FPP in the biosynthesis of certain sesquiterpene types, and it further suggests a universal role for *trans; trans*-FPP in sesquiterpene cyclizations. Thus, in those instances where direct cyclization is not stereochemically possible, preliminary conversion to a nerolidyl intermediate could be proposed, analgous to the linaloyl intermediate suggested in the monoterpene series. Nerolidyl intermediates in sesquiterpene cyclization processes have been suggested previously (Arigoni 1975; Andersen et al. 1978), and the biosynthesis of the cyclopentanoid sesquiterpene cyclonerodiol in cell-free preparations of the fungus *Giberella fujikori* was recently shown to involve free nerolidyl phyrophosphate (derived from all *trans* FPP) as an intermediate (Cane 1980a,b). The cyclization is, however, atypical of higher plant sesquiterpenes, in that C(1) of the acyclic precursor is not involved in the formation of the new C-C bond. Thus, the general involvement in sesquiterpene biosynthesis of free nerolidyl pyrophosphate (or of the enzyme-bound nerolidyl cation suggested here) cannot yet be assessed. In any case, there no longer appears to be any compelling reason to regard the conversion of *trans, trans*-FPP to *cis, trans*-FPP as a precondition for cyclization in the sesquiterpene series. Thorough discussion of the question of acyclic sesquiterpene precursors can be found in recent reviews by Cane (1980a,b).

The exact nature of the acyclic precursor notwithstanding, a very comprehensive and unified biogenetic scheme for the formation of sesquiterpenes has evolved, based on the original isoprene rule (Ruzicka et al. 1953) and extended through detailed consideration of steric and electronic factors (Hendrickson 1959; Parker et al. 1967; Rucker 1973). As with other biogenetic outlines, the proposed transformations in the sesquiterpene series are supported by the structural relationships of co-occurring metabolites, plausible chemical conversions and *in vivo* tracer studies [see for example Rucker (1973); Arigoni (1975); Cordell (1976); Andersen et al. (1978) and Cane (1980b)]. The basic scheme for the formation of a few selected skeletal types is illustrated in Figure 11, but with the substitution of enzyme-bound nerolidyl species derived from FPP as precursors (rather than the *cis*-Δ^2 and *trans*-Δ^2 cations as is often done). Cyclization can be seen to involve electrophillic attack by the incipient C(1) cation on either the central or distal double bond of the C_{15} precursor, the stereochemical outcome being determined by the configuration at the enzyme-bound carbon, the conformer cyclized and the orientation of the isopropylidene group. As was the case for the monoterpene series (Fig. 6), the possibilities for generating enantiomeric compounds in the sesquiterpene series are both numerous and self-evident. On completion of the initial ring closure, the reaction may be terminated by addition of a nucleophile (e.g., $^-$OH) or by loss of a proton with double bond formation, leading to the parent alcohol or olefin. Alternatively, internal addition via the remaining double bonds of the neutral species or a cationic equivalent may occur, along with rearrangements, giving rise to various bicyclic and tricyclic skeletal derivatives. Such schemes resemble those proposed in the monoterpene series, and virtually all common monoterpenes have sesquiterpenes analogs (Fig. 12). However, the additional C_5 unit of the sesquiterpenes and the ability to form 10- and 11-membered ring intermediates affords far greater flexibility in the generation of C_{15} skeleta than is possible for the C_{10} series.

While it is theoretically possible to derive essentially all sesquiterpenes from FPP, relatively few of the biogenetic schemes have been experimentally tested in higher plants. Even fewer cyclizations have been examined at the molecular level with plant systems. Thus, cell-free preparations from *Kadsura japonica* convert FPP to germacrene C (Fig. 2) (Morikawa et al., 1971) and extracts of *Carum carvi* convert FPP to germacrene D (Fig. 2) (Croteau, 1980b). Germacrene D is the macrocyclic analog of the monoterpene phellandrene (Fig. 1), however the sesquiterpene cyclase was unable to efficiently synthesize β-phellandrene from GPP. Similarly, monoterpene cyclases do not efficiently synthesize sesquiterpene analogs (e.g., α-pinene synthetase from *Salvia officinalis* cyclizes GPP to α-pinene fifty times faster than it cyclizes FPP to α-bergamotene, while γ-terpinene synthetase from *Thymus vulgaris* converts GPP to γ-terpinene thirty times faster than it converts FPP to β-curcumene) (Croteau, 1980b). Thus, it appears unlikely that the same cyclase enzyme synthesizes monoterpene and sesquiterpene analogs, a suggestion supported by the general lack of co-occurrence of such analogous compounds. On the other hand, it is quite likely that the mechanisms involved in the biosynthesis of analogous mono- and sesquiterpenes are similar, but comparative studies with monoterpene cyclases and the corresponding sesquiterpene cyclases will be required to explore this point.

Most recent advances in the area of sesquiterpene biosynthesis have been made using *in vivo* systems with the isotope tracer approach employing doubly radio labelled precursors (e.g., [$5-^3H, 2-^{14}C$-MVA]) and multiply ^{13}C-labelled precursors (e.g., [$1,2-^{13}C_2$] acetate.) The latter approach has been particularly useful in the study of sesquiterpene biosynthesis, but has been limited, for the most part, to investigations of fungal metabolites and phytoalexins where rather high incorporation rates have been achieved (Cordell, 1976; Stoessel et al., 1977, 1978; Cane, 1980b). The technique has been successfully applied to the biosynthesis of γ-bisabolene and paniculide B in suspension cultures of *Andrographis paniculata* (Overton and Picken, 1976). In this instance, the ^{13}C NMR spectrum of paniculide B derived from [$1,2-^{13}C_2$] acetate exhibited six pairs of coupled doublets and three enhanced singlets [C(4), C(8), C(12)], allowing the

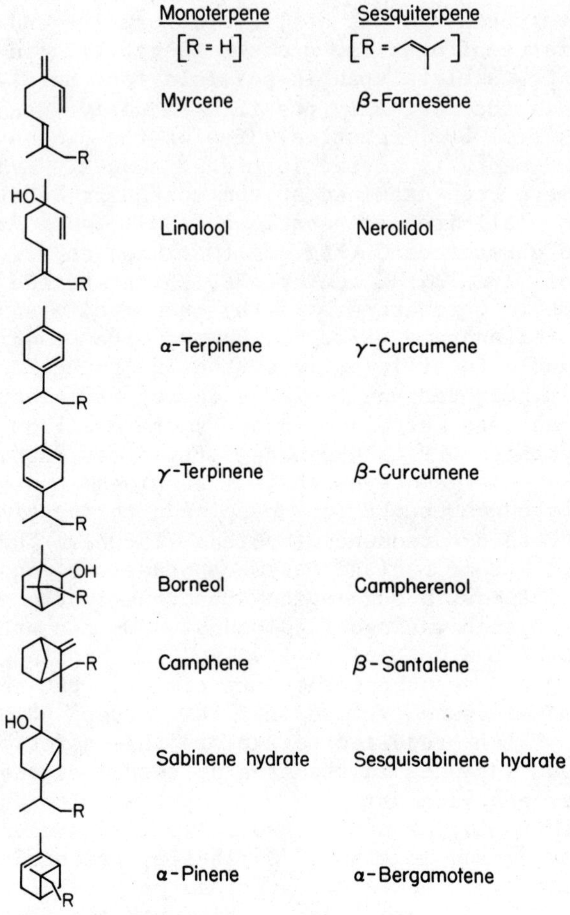

Monoterpene [R = H]	Sesquiterpene [R = ⁓Y]
Myrcene	β-Farnesene
Linalool	Nerolidol
α-Terpinene	γ-Curcumene
γ-Terpinene	β-Curcumene
Borneol	Campherenol
Camphene	β-Santalene
Sabinene hydrate	Sesquisabinene hydrate
α-Pinene	α-Bergamotene

Fig. 12. Some monoterpenes and their sesquiterpene analogs.

pathway to the lactone, via γ-bisabolene of known configuration, to be deduced (Fig. 13). More common than the ^{13}C NMR approach to the study of higher plant sesquiterpenes has been the use of ^{3}H-, ^{14}C-doubly labelled precursors. Particularly noteworthy have been studies on the origin of d-longifolene in *Pinus* (Arigoni 1975). Chemical degradation of longifolene derived from [2-^{14}C]MVA established a labelling pattern consistent with the pathway in Figure 14. The product isolated when [5-^{3}H,2-^{14}C] MVA was the substrate contained all six possible tritium atoms, which, when subsequently located, supported the pathway proposed and demonstrated the indicated 1,3-hydride shift in the macrocyclic intermediate (Fig. 14). Furthermore, the retention of all tritium atoms of the precursor ruled out oxidative transformation of *trans, trans*-FPP to *cis, trans*-FPP prior to cyclization (i.e., tritium at C(1) of FPP derived from [5-^{3}H]MVA was not lost in the conversion of the *trans*- Δ^2-double bond of this acyclic precursor to the corresponding *cis*-double bond of the humulane intermediate). That longifolene was in fact generated from *trans, trans*-FPP was confirmed by the finding that longifolene derived from 4R-[4-^{3}H]MVA contained all three possible tritium atoms. Had *cis, trans*-FPP been formed directly as the precursor, tritium would, presumably, have been lost in the *cis*-condensation of IPP. 5R-[5-^{3}H,2-^{14}C]- and 5S-[5-^{3}H,2-^{14}C]MVA, as well as the aforementioned 4R-[4-^{3}H]MVA, were then employed to examine stereochemical features of the 1,3-hydride shift. These studies on the origin of d-longifolene, and similar studies of the biosynthesis of l-longifolene and several cadalane-type sesquiterpenes in fungal systems, lead Arigoni (1975) to propose a unified theory relating the cyclization of FPP to the stereospecificity of the 1,3-hydride shift and the absolute configuration of the resulting product.

Another example of the utility of 4R-[4-^{3}H,2-^{14}C]MVA as a sesquiterpene precursor is provided by the studies of Brooks and associates (Zabkiewicz et al. 1969; Brooks and Keates 1972) which supported the involvement a 1,2-hydride shift and methyl migration in the transformation of a eudesmane intermediate (Fig. 11) to the eremophilane sesquiterpene petasin in *Petasites hydridus* (Fig. 15). The results were also consistent with the intermediacy of *trans, trans*-FPP in the pathway.

c. *Diterpenes*. In spite of the great structural diversity of the diterpenes, meaningful classification of these compounds in biogenetic terms is possible as was the case for mono- and sesquiterpenes. The vast majority of the diterpenes are cyclic, containing from one to five rings and comprising many skeletal modifications. As described above, in the mono- and sesquiterpene series, the most characteristic apparent mode of cyclization involves loss of the terminal pyrophosphate with ring closure via a double bond, leading, for example, to the *p*-menthane (C_{10}) or germacrane (C_{15}) skeleton and

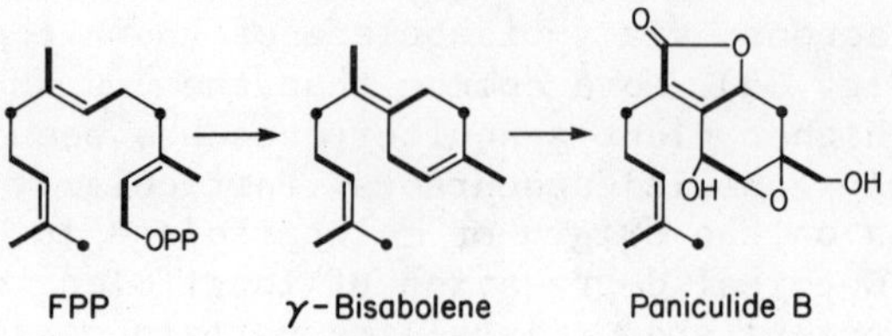

Fig. 13. Labeling pattern of paniculide B derived from $[1,2-^{13}C_2]$ acetate in Andrographis paniculata and determined by ^{13}C NMR. Heavy lines indicate coupled doublets, while closed circles indicate enhanced singlets corresponding to C(4), C(8) and C(12) of farnesyl pyrophosphate (after Overton and Picken 1976).

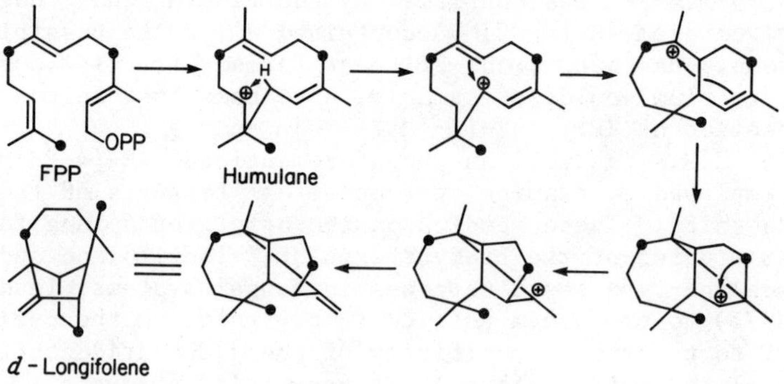

Fig. 14. Proposed pathway for the conversion of farnesyl pyrophosphate to d-longifolene via a humulane intermediate. Closed circles indicate the position of labeling from $[2-^{14}C]$mevalonate (after Arigoni 1975).

Fig. 15. Proposed formation of the eremophilane skeleton of petasin from a eudesmane-type intermediate (after Zabkiewicz et al. 1969 and Brooks and Keates 1972). R is the 2-methyl-2-butenoyl group.

to further cyclizations (e.g., bornane and eudesmane types).
Only a few cases of this mode of cyclization of GGPP appear to
occur in the diterpene series, typical examples being the
diterpene macrocyclics of the cembrene-type (Dauben et al.
1965; Birch et al. 1972) and casbene (Robinson and West
1970a,b; Sitton and West 1975), the latter which contains
a cyclopropane ring in addition to the macrocyclic ring.
This group of diterpenes is considered to be biogenetically
related through a similar macrocyclic precursor (Fig. 16)
(West 1980), and may originate by mechanisms similar to
those involved in the construction of analogous mono- and sesquiterpenes (Cf., limonene to cembrene A, and bicyclogermacrene to
casbene). Further cyclization of the macrocyclic ring (e.g., by a
process similar to those involved in the formation of pinenes and
bergamotenes of the C_{10} and C_{15} series) can lead to a carbon
skeleton of the verticillol-type (Erdtman et al. 1964) (Fig. 16).

The origin of casbene, which is thought to be a phytoalexin in
castor bean (Sitton and West 1975), has been examined by Robinson
and West (1970a, b). The enzyme responsible for the cyclization
of GGPP to casbene is localized in proplastids in castor bean
tissue (West et al. 1979), and the casbene synthetase has been
partially purified and characterized (Dueber et al. 1978). The
enzyme exhibits a M.W. of ~53,000, requires a divalent cation for
activity (Mg^{++}) and possesses other properties similar to
monoterpene synthetases and other diterpene synthetases (see
below).

The major mode of cyclization in the diterpene class
appears to involve the initial protonation of the isopropylidene
double bond of GGPP, followed by several distinct ring closure
sequences (Fig. 17). Thus, GGPP initially cyclizes to a bicyclic
labdane intermediate (Labda-8(17), 13-dien-15-yl pyrophosphate)
possessing either of two possible antipodal A/B ring junctions.
Diterpenes of both the "normal" and "enantiomeric" (ent-)
series are known [both antipodes of labda-8(17), 13-dien-15-oic
acid have, in fact, been isolated from the same species (Bevan
et al. 1968)], the stereomer formed via such antiparallel
additions presumably depending on the conformer of GGPP maintained
at the enzyme surface (Scott et al. 1964). Many derivatives
of the labdane skeletal type (e.g., kolavenol (Fig. 3)] are
known (Devon and Scott 1972). The next cyclization step involves
the loss of the pyrophosphate moiety with ring closure to yield the
tricyclic pimarane nucleus (Fig. 17). In this instance also,
cyclization can result in enantiomeric compounds (at the methyl:
vinyl substituted carbon), and again both series are naturally
occurring and reflect the stereoselectivity of the cyclizing enzyme
[Cf. rimuene (Fig. 3) and sandaracopimaradiene (Fig. 18)]. The
pimarane intermediate may serve as the progenitor of several other
tricyclic diterpene types, including the abietanes, rosanes, and
cassanes (Fig. 17), or, by further cyclization, lead to polycyclic

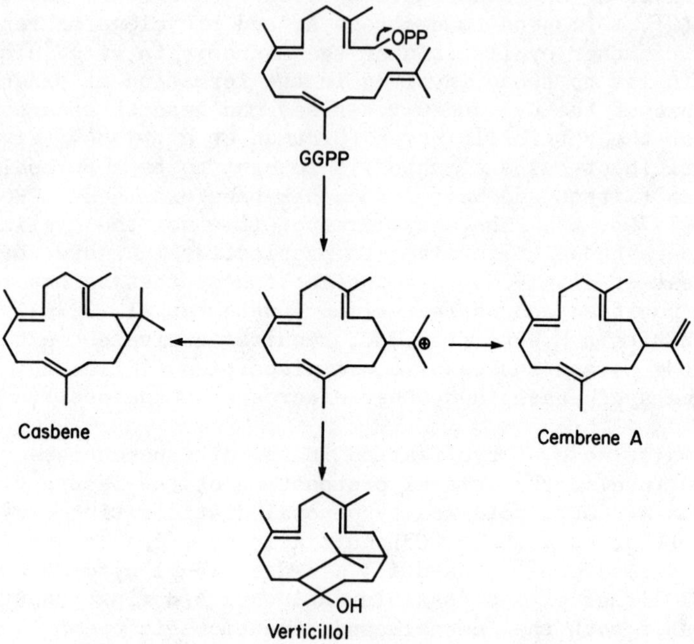

Fig. 16. Proposed formation of macrocyclic diterpenes from geranylgeranyl pyrophosphate (adapted from West 1980b).

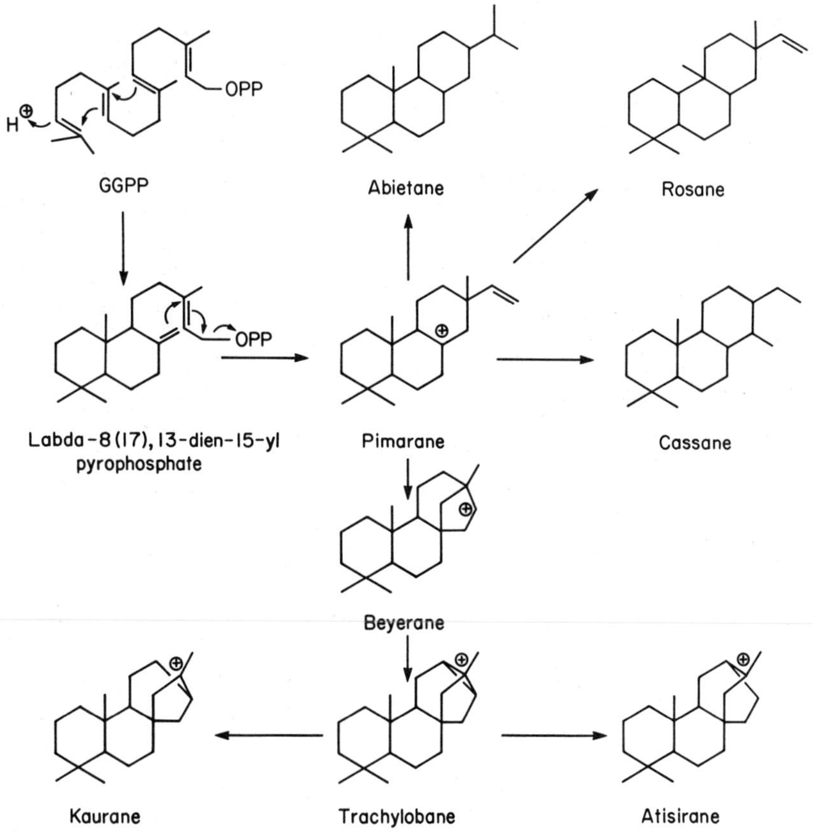

Fig. 17. Proposed cyclixation of geranylgeranyl pyrophosphate to polycyclic diterpenes.

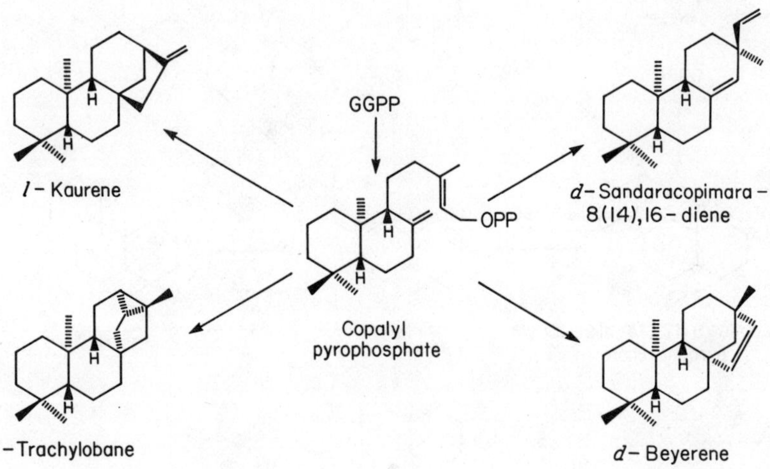

Fig. 18. Formation of *Ricinus communis* diterpenes by cyclization of geranylgeranyl pyrophosphate through copalyl pyrophosphate (after Shechter and West 1969 and Robinson and West 1970 a,b).

diterpenes of the beyerane, trachylobane, kaurane and atisirane type (Fig. 17). The cyclization scheme outlined here and described in detail elsewhere (Wenkert 1955; Oehlschlager and Ourisson 1967; Hanson 1971; West 1980b) is based on bigenetic consideration of naturally occurring (often co-occurring) diterpenes (Briggs and White 1975) and on quiet reasonable chemical transformations (Oehlschlager and Ourisson 1967; Hanson 1972; Nakaniski 1974), and it is supported by *in vivo* labelling studies [primarily with fungal systems; see Oehlschlager and Ourisson (1967) and West (1980b) for lead refs.].

Relatively little is known about the construction of diterpene skeleta at the enzyme level, as few cell-free preparations capable of carrying out the requisite cyclizations are available. Cell-free extracts of castor bean *(Ricinus communis)* simultaneously convert GGPP to four different entpolycyclic diterpene hydrocarbons (d-sandaracopimara-8(14), 16-diene, d-beyerene, l-trachylobane and l-kaurene) as well as to the aforementioned macrocyclic diterpene casbene (Fig. 18) (Shechter and West 1969; Robinson and West 1970a,b). Each of the polycyclic diterpenes is synthesized through the intermediary of *ent-l-*labda-8(17),13-dien-15-yl pyrophosphate (copalyl pyrophosphate) that is generated from the cyclization of GGPP by a copalyl pyrophosphate synthetase (the A enzyme activity). d-Beyerene, l-trachylobane and l-kaurene are biogenetically related as indicated in Figure 17 and they are presumed to be synthesized from copalyl pyrophosphate by closely related cyclizing enzymes (the B enzyme activities). Thus, the overall reaction starting from GGPP is considered as the summation of the two enzymatic activities (the AB activity). d-Sandaracopimara-8(14),16-diene has the opposite configuration at C(13) (i.e., the methyl:vinyl substituted carbon) from the expected intermediate of the other polycyclic compounds and would, thus, be expected to arise from copalyl pyrophosphate via a different type of cyclase (B enzyme). To examine this question, West and associates (Robinson and West 1970b; Simcox 1976; West 1980b) purified the cyclization enzymes from castor bean and found, in addition to a peak of AB activities capable of synthesizing all four products from GGPP, two other regions containing only B type activity. One specifically converted copalyl pyrophosphate to l-kaurene, and the other specifically converted copalyl pyrophosphate to d-sandaracopimaradiene, thus suggesting that distinct B activities catalyzing the synthesis of single end-products do exist. B-type enzymes, capable of converting copalyl pyrophosphate to kaurene, have also been isolated (free of A-type activity) from other sources (Yafin and Shechter 1975), and recently an A-type enzyme (GGPP to copalyl pyrophosphate) has been separated from the B enzyme of *Marah macrocarpus* (Duncan 1980; see also West 1980b). The availability of separated

A and B enzymes should allow more detailed study of the mechanism of diterpene synthesis.

In their gross properties, the diterpene synthetases of higher plants resemble the various monoterpene synthetases described earlier. For example, kaurene synthetase from *Marah macrocarpus* has a M.W. of 82,000, exhibits a pH optimum at 6.6, requires Mg^{++} for activity, has a K_m for the prenyl pyrophosphate substrate in the μM range, and is inhibited by thiol directed reagents (Upper and West 1967; Frost and West 1977; West 1980b), all properties very similar to those of γ-terpenine synthetase from *Thymus vulgaris* (Poulose and Croteau 1978b). Among the diterpene synthetases, kaurene synthetase has received the most experimental attention because kaurene is the precursor of the gibberellin plant hormones. West (West et al. 1979; West 1980b) has compared the properties of kaurene synthetases (A and B activities) from various plant sources. The conversion of l-kaurene to the gibberellins, while outside the scope of this review, has been described in detail elsewhere (Hedden et al. 1978; Graebe and Ropers 1978; West 1980b).

4.3 *Secondary Transformations*

While relatively few "cyclases" are thought to determine the basic structures of the mono-, sesqui-, and diterpenes, perusal of any compendium of terpene compounds (Devon and Scott 1972) will illustrate the very large number of derivatives of each skeletal type found in nature, and give some appreciation of the large number of secondary enzymatic transformations presumed to occur among the various terpenoid classes. Simple hydrocarbons and highly oxygenated terpenes are known, nitrogen, halogen and even sulfur containing terpenes are encountered, and the number of positional, geometric and stereo isomers is legion. It is clearly the secondary transformations that provide the great number and diversity of the terpenes, and it is these modifications that are generally responsible for the biological function of many of these compounds.

While the metabolism of the terpenes encompasses a diverse assortment of biochemical transformations, the higher plant terpenes, especially those isolated from the essential oils and resins, are commonly hydrocarbons or simple oxygenated derivatives and their conjugates (alcohols, aldehydes, ketones, acids, esters, glycosides, etc.). It is commonly believed that the initial cyclization (with or without accompanying rearrangement) leads to the parent skeletal compound (generally at the olefin or alcohol level), which is subsequently modified by oxygenation, dehydrogenation, isomerization or conjugation reactions. This is somewhat of an oversimplification, as among the sesquiterpene and diterpene series further rearrangements of the parent carbon

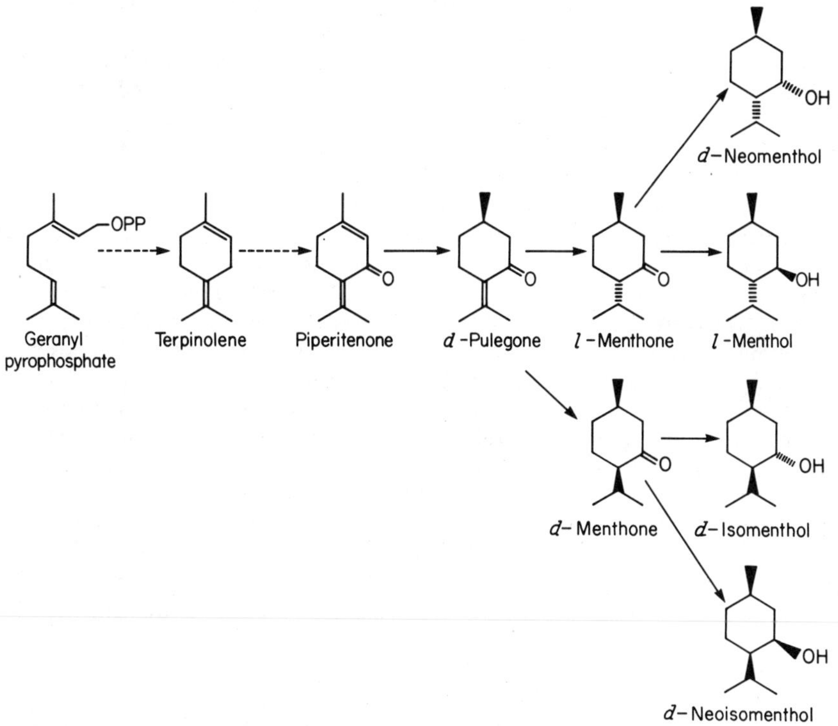

Fig. 19. Pathyway for the formation of menthol stereomers in *Mentha piperita* (after Loomis 1967). Broken arrows indicate postulated reactions.

skeleton may occur after oxidative modification [e.g., in the conversion of the kaurane to the gibberellane skeleton by ring contraction (MacMillan 1971)].

Early speculations concerning terpene metabolism were based largely on the co-occurrence of structurally related compounds and on their temporal variations, and on chemical reasoning often supported by biomimetic syntheses. *In vivo* studies employing isotopically labelled basic precursors such as acetate and MVA, have been utilized to deduce probable pathways via the determination of labelling patterns and the time-course of appearance of products and intermediates. The direct tracer approach, involving the administration of suspected precursors and intermediates, has also proved effective in defining metabolic sequences. More recently, cell-free systems capable of sustaining various terpene modifications have been isolated, permitting exploration of terpene metabolism at the enzyme level.

The most thoroughly studied of any metabolic sequence in the monoterpene series is that involved in the formation of menthol stereomers (Fig. 19). Investigations with peppermint *(Mentha piperita)*, primarily by Loomis and associates (Battaile and Loomis 1961; Burbott and Loomis 1967, 1969; Battaile, Burbott and Loomis 1968), have confirmed this scheme through several types of evidence. Analysis of the monoterpenes from different leaves of mint showed a predominance of piperitenone and pulegone in the youngest leaves, with a trend toward menthones, then menthols, in the oldest leaves. When $^{14}CO_2$ was administered to the plants, the unsaturated ketones were labelled first, followed by menthones, and later menthols. Most of the reactions were also demonstrated directly by feeding labelled monoterpenes to peppermint leaves. Loomis was one of the first to take into account stereochemical considerations in biosynthetic conversion of piperitenone to the menthol isomers, and this group has recently demonstrated, in cell-free preparations, several of the stereospecific, NADPH-dependent reductases involved in this sequence of reactions (Burbott and Loomis 1980). The origin of piperitenone is still uncertain, and this proposed conversion is illustrated by broken arrows in Figure 19. The conversion of acyclic precursors to terpinolene has been demonstrated (Croteau and Karp 1976a), but the mode of oxygenation of terpinolene to piperitenone has not been determined.

Another family of monoterpenes that has recently been examined is the aromatic monoterpenes. Time-course studies and direct incorporation studies were utilized to demonstrate γ-terpinene to ρ-cymene and the hydroxylation of this intermediate to thymol in *Thymus vulgaris* (Poulose and Croteau 1978a) (Fig.20). The requisite enzymes have been demonstrated in cell-free systems, but only the γ-terpinene synthetase has been examined in any detail (Poulose and

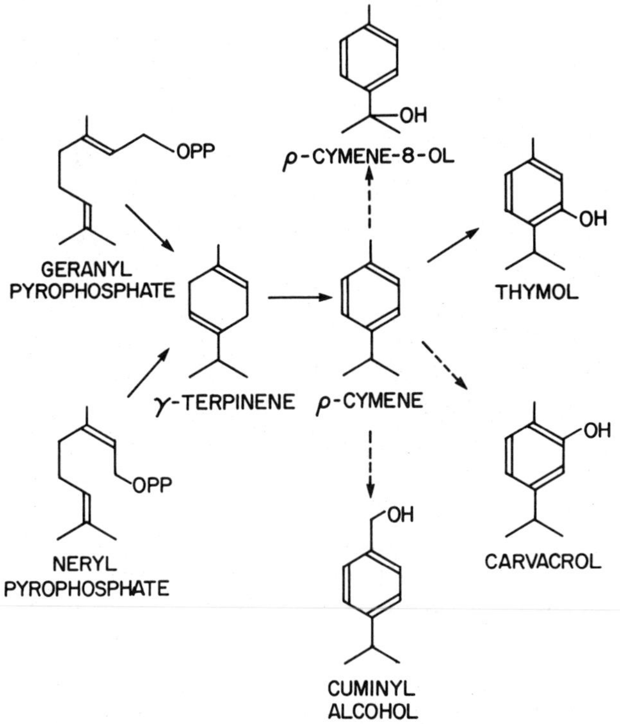

Fig. 20. Proposed general pathway for the biosynthesis of aromatic monoterpenes. Solid arrows indicate reactions demonstrated either *in vivo* or *in vitro* while broken arrows indicate postulated steps.

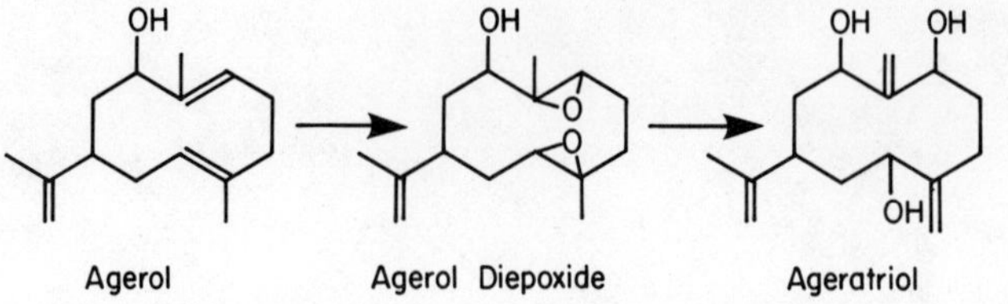

Fig. 21. Conversion of agerol to agerol diepoxide and ageratriol (after Bellesia et al. 1975).

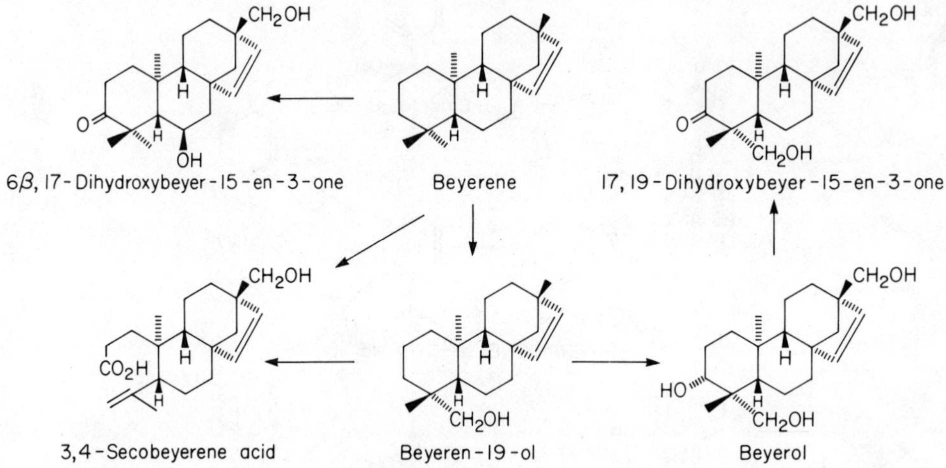

Fig. 22. Diterpene transformations in *Beyeria leschenaultii* (adapted from Bakker et al. 1972).

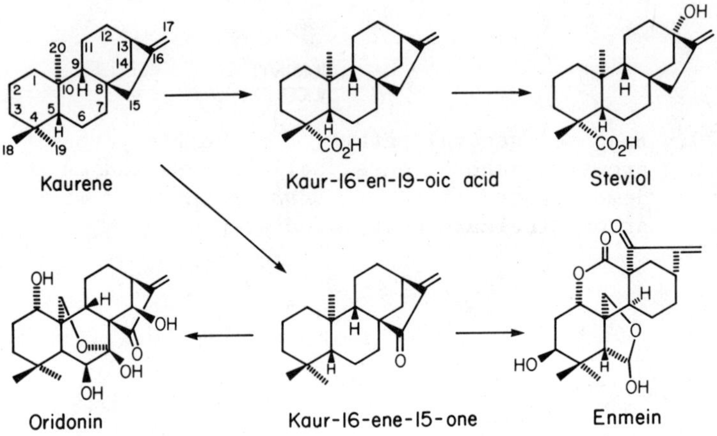

Fig. 23. Conversion of kaurene to steviol (after Bennett et al. 1967, and Hanson and White 1968) and to enmein and oridonin (after Fujita et al. 1976).

Croteau 1978b). Unlike the γ-terpinene synthetase, the other enzymes of the metabolic sequences are membrane-associated. More detailed discussion of monoterpene metabolism can be found elsewhere (Croteau 1980a,b).

The sesquiterpene and diterpene metabolites of higher plants are of types similar to those of the monoterpene series, although polyoxygenated compounds are more frequently encountered among the former two classes. As was the case with monoterpenes, oxygen may be introduced in some instances during the cyclization [as predicted for bisabolol and hedycaryol (Fig. 2), for example]. For the most part, however, oxygen appears to be introduced via allylic, methylene, or methyl hydroxylation, or by epoxidation or hydration of double bonds (Nes and McKean 1977). Dehydrogenases likely convert the alcohols to the corresponding ketones, aldehydes and carboxylic acids. Such derivatives undergo further transformations, as furans and lactones are numerous (Sutherland and Park 1967; Roberts 1972; Hanson 1972; Geissman 1973). Aromatic rings are encountered among the sesquiterpenes [e.g., cuparene, α-curcumene (Fig. 2)] and aromatization process may be similar to that involved in the biosynthesis of the monoterpene ρ-cymene.

While the analysis of co-occurring metabolites, time-course studies, and biomimetic simulations have often aided in elucidating metabolic sequences among the sequi- and diterpenes (Geissman 1973; Paknikar et al. 1975; Martin et al. 1976; Herz 1977; Stoessl et al. 1977; Andersen et al. 1978), the more direct *in vivo* approach involving the feeding of suspected intermediates has also been carried out. For example, the direct conversion of the sesquiterpene agerol to ageroldiepoxide and ageratriol was established (Fig. 21) (Bellesia et al. 1975). Similar studies have been carried out to elucidate secondary transformations in the diterpene series. Thus, Bakker and Coworkers (1972) utilized labelled *ent-d*-beyerene, beyeren-19-ol and beyerol to establish biosynthetic relationships among the diterpenes of *Beyeria leschenaultii*, and to examine stereoselectivity in the formation of a diterpene seco-acid (Fig. 22). Direct feeding experiments have also been employed to demonstrate that *ent-l* -kaurene and *ent*-kaur-16-en-19-oic acid are precursors of steviol (Bennett et al. 1967; Hanson and White 1968), and that *ent-l*-kaurene and *ent*-kaur-16-en-15-one are progenitors of the highly oxygenated diterpene bitter principles enmein and oridonin (Fujita et al. 1976) (Fig. 23). As with the examples above, most cyclic diterpenes that have been examined thus far are products of further enzymatic modification of parent cyclic compounds such as beyerene and kaurene. Perhaps the most extensively studied secondary transformation of any terpene (other than those of the steroid series) is the conversion of *ent-l*-kaurene to the gibberellin plant hormones. Although not strictly related to the diterpenes under consideration here, the oxidative transformations of kaurene (Hedden et al. 1978; Graebe and Ropers 1978) are probably typical of diterpenes in general, and the various oxygenases and dehydrogenases involved in such transformations are probably representative of classes of enzymes involved in the secondary transformations of other terpene types (Graebe and Hedden 1974; West 1980a).

Unlike the rather unique enzymes involved in the basic construction of terpenoids, the many classes of enzymes involved in the secondary transformations of terpenes are of more common biochemical type (mixed-function oxidases, dehydrogenases, etc.) and it is not possible to review many specific examples here. However, an important general question concerning the enzymes of terpenoid metabolism is whether such secondary transformations are catalyzed by commonly occurring enzymes of low specificity, or by unique, highly specific enzymes. This question has been a matter of considerable speculation (Loomis 1967; Banthorpe et al. 1972a; Croteau and Loomis 1975), but relatively little information is available except from transformations among the monoterpene series. From this work, it would appear that the enzymes of monoterpene metabolism exhibit a significant degree of structural specificity. For example, a geraniol dehydrogenase has been isolated from *Citrus* spp. and resolved from ethanol dehydrogenase (Potty and Bruemmer 1970a), and a reductase from this tissue reduces the exocyclic double bond of d-limonene but not that of l-limonene (Potty and Bruemmer 1970b). Similarly, several positionally-specific double-bond reductases are involved in piperiterone metabolism in peppermint (Fig. 19) (Burbott and Loomis 1980). These NADPH-dependent enzymes act only on members of the C(3) oxygenated piperitenone series; members of the C(2) oxygenated carvone series are not substrates. The monoterpenol dehydrogenase from sage that oxidized borneol to camphor has been partially purified and characterized in detail (Croteau et al. 1978). Several types of evidence indicate that the same dehydrogenase, which is distinct from alcohol dehydrogenase, catalyzed the key oxidations of d-borneol to d-camphor and of l-3-neoisothujanol to l-3-isothujone (Fig. 1) [d-camphor and l-3-isothujone are major monoterpenes of sage (Lawrence et al. 1971)]. The enzyme does not, however, readily oxidize stereomers of d-borneol and l-3-neoisothujanol, or other monoterpenols, suggesting that this dehydrogenase is specific for the particular group of monoterpenols produced by sage. Similar "group specific" monoterpenol dehydrogenases were recently isolated from fennel (*Foeniculum vulgare*) and *Tanacetum vulgare* (Croteau and Felton 1980a), thereby extending this concept to other plant families. Such group-specific enzymes might also be involved in other types of terpene secondary transformations.

5. CATABOLISM

The traditional view has held that terpenes of the essential oils and resins are stable end-products of metabolism (Paech 1950; Sandermann 1962; Mothes 1973); however, considerable evidence now indicates that monoterpenes, sesquiterpenes and diterpenes are not simply accumulated, but that they are metabolically active. In the case of monoterpenes, this conclusion is based on many different types of evidence including analyses of both short-term (diurnal) and long-term variation in monoterpene content (Burbott and Loomis 1967; Schröder 1969; Weiss and Flück 1970; Dro and Hefendehl 1973; Adams and Hagerman 1977). For example, rapid and permanent turnover of monoterpenes was shown to occur in flowering peppermint plants, and during this period of active catabolism otherwise undamaged oil glands appeared to be emptied of their contents (Fig. 24a) (Burbott and Loomis 1967; Croteau and Martinkus 1979). Additionally, the time-course of incorporation of labelled precursors such as MVA, CO_2 and glucose into monoterpenes showed that label passes rapidly through the monoterpene pool without changing the pool size (Scora and Mann 1967; Hefendehl et al. 1976; Burbott and Loomis 1969; Francis and O'Connell 1969; Croteau and Loomis 1972). In peppermint cuttings, nearly 40% of the monoterpenes derived from $^{14}CO_2$ are catabolized within 10 hours of maximal labelling (Fig. 24b) (Croteau et al. 1972a). Similarly, exogenous labelled monoterpenes are rapidly degraded by plants. For example, [^{14}C] terpinen-4-ol (Fig. 1) when administered to *Tanacetum vulgare* cuttings is rapidly converted into water-soluble materials, including amino acids and sugars (Banthorpe and Wirz-Justice 1969; Banthorpe et al. 1972a), thus implying oxidative degration to small molecules followed by resynthesis to primary metabolites.

As monoterpenes and sesquiterpenes appear to be synthesized at different sites, the kinetics of turnover of these terpene classes might be expected to differ. Such differences have been observed in tracer studies (Croteau and Loomis 1972; Croteau et al. 1972a; Banthorpe and Ekundayo 1976; Regnier et al. 1968). Similar studies have shown that the biological half-life of diterpenes in *Marrubium vulgare* is about 24 hours (Breccio and Badiello 1967). The degradation of [^{14}C]sclareol (Fig. 4) was demonstrated directly when this diterpene was administered to *Salvia sclarea* (Nicholas 1964). Thus, many types of evidence reviewed in detail elsewhere (Loomis 1967; Loomis and Croteau 1973; Gleizes 1976), provide compelling arguments that terpenes do undergo natural turnover in plants and that plants are capable of recycling terpenes into primary metabolites. The onset and rate of terpene turnover appear to be highly dependent on environmental factors and on the physiological and/or developmental status of the plant, and

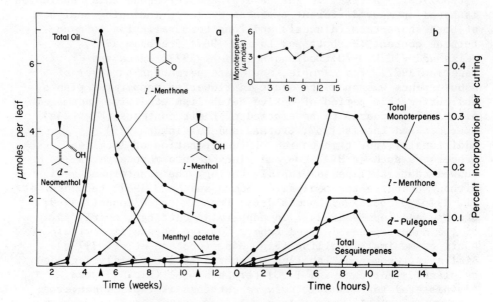

Fig. 24. (a) Monoterpene composition of midstem peppermint leaves as a function of development. The first arrow indicates the approximate time of floral initiation and the second arrow indicates the approximate time of full bloom (from Croteau and Martinkus 1979). (b) Time-course of labeling of pepperming mono- and sesquiterpenes after 1 hr exposure to $^{14}CO_2$ in the light (from Croteau et al. 1972a).

present evidence suggests that synthesis, storage and catabolism of terpenes may be controlled by the balance between photosynthesis and the utilization of photosynthate, or by the growth-differentiation balance (Burbott and Loomis 1967; Croteau et al. 1972b; Loomis and Croteau 1973).

Preliminary studies on monoterpene catabolism have been carried out with peppermint. In this species catabolism is apparently accelerated near the onset of flowering, coincident with the conversion of methone to menthol and to lesser quantities of menthyl acetate and neomenthol (Burbott and Loomis 1969; Croteau and Martinkus 1979) (Fig. 24a). Detailed studies of the metabolism of l-[G-^{3}H] menthone in peppermint confirmed earlier observations that menthone was converted to menthol and menthyl acetate, and furthermore revealed that a significant proportion of the menthone was transformed to d-neomenthyl-β-D-glucoside (Croteau and Martinkus 1979). Little neomenthyl acetate or menthyl-β-D-glucoside was formed from [G-^{3}H]menthone, indicating a high degree of specificity in the metabolic disposition of the epimeric reduction products of the ketone (i.e., menthol and menthyl acetate accumulate in the volatile oil, while neomenthol is specifically converted to the water-soluble glucoside). Further studies have suggested that the specificity observed is a result of compartmentation of each stereospecific dehydrogenase with the appropriate auxiliary enzyme (i.e., the transacetylase or the transglycosylase), but the nature of the compartments has yet to be determined (Croteau 1980b; Martinkus-Taylor and Croteau 1980). Unlike most enzymes of monoterpene metabolism, the transacetylase and transglucosylase involved in this "catabolic sequence" are notable for their lack of substrate specificity (Croteau and Hooper 1978; Martinkus-Taylor and Croteau 1980).

The neomenthol glucoside formed in peppermint leaves appears to be transported to the roots and partially degraded to unidentified water-soluble products that are no longer glucosylated (Martinkus-Taylor and Croteau 1980). This observation suggests that monoterpenes may be degraded at sites other than the glandular trichomes, and it supports suggestions (Francis 1971; Skopp and Hörtser 1976; Hörster 1979) that glycosides are the transport derivatives of terpenes. If these suggestions are correct, then glycosylation of terpenols may be an essential first step in terpene catabolism. Although the catabolism of terpenes may have great physiological and ecological significance, virtually nothing is known about the regulation of the process or about the pathways or mechanisms of such catabolism.

6. BIOLOGICAL SIGNIFICANCE

Terpenoid substances are commonly synthesized in response to specific developmental and ecological needs of the plant (Swain 1974). Thus, for example, the terpenoid plant hormones, abscisic acid and the gibberellins, have a clear physiological function, while an equally obvious ecological role is attributed to the terpenoid phytoalexins, such as rishitin (Kúc and Lister 1978) and casbene (Sitton and West 1975), synthesized in response to fungal infection. For the vast majority of trichomal terpenes no obvious physiological role has been identified, and the sequestering of terpene formation within the trichomes would suggest that such a role is unlikely. Yet, the previously described modulation of terpene metabolism with development, and the observed catabolism of trichomal terpenes, imply that a physiological function, as yet obscure, does indeed exist.

While the role of glandular terpenes *within* the plant is uncertain, the trichomes are well-situated for a communicative and defensive function, and the terpenes contained therein well-suited for such diverse ecological purposes. Thus, volatile terpenoids provide the ability to influence another organism at some distance from the source plant, while complexity of composition and structure of the constituent mixture confers the ability to transmit a very selective biological message. A lipophilic nature allows for some persistence in the largely aqueous biosphere, and the unusual structural features of these plant products promote a degree of biological stability, with possible toxic consequences for other life forms.

A number of ecological interactions mediated by trichomal terpenes are well-documented. The competitive advantage provided by monoterpene phytotoxins of *Salvia* species is a noteworthy example (Müller 1971; Müller and Chou 1972) as is the highly selective, sesquiterpene-dependent pollination strategy of *Ophrys* orchids (Kullerberg and Bergström 1975). Trichome-derived terpenoids may provide a means of repelling herbivores during vegetative growth, while attracting them later as a tactic in seed dispersal. Such adaptational interactions can be quite complex, yet subtle and easily overlooked. Thus, the ecological significance of trichomes and their highly diverse terpenoid constituents might well be underestimated.

7. CONCLUSIONS

Progress in understanding the biochemistry of the terpenoids produced in glandular trichomes has been impeded by the nature of both the terpenes and the plants that produce them. The plant terpenes generally occur as complex, often unstable, mixtures that are difficult to purify by classical techniques. Compartmentation effects in higher plants commonly result in rather low incorporation rates of exogenous terpenoid precursors, thus limiting the amount of information available from *in vivo* studies. Similarly, the enzyme level approach has been hampered by difficulties associated with obtaining active, stable cell-free preparations from intractable plant sources. In spite of the problems, progress in the biochemistry of plant terpenes is being made as illustrated by the examples described herein. Application of increasingly sophisticated chemical and biochemical approaches will result in continued progress, particularly at the enzyme level where the fundamental questions lie. Cyclization processes have received the most attention, while other areas are wide open to exploration. Catabolism, for instance, has only recently been explored, primarily in the monoterpene series, and regulatory aspects have thus far received little attention, except in the diterpene series. Many aspects of the relationship of terpene biochemistry to glandular trichome structure remain to be elucidated, and, while the terpenes are no longer regarded as metabolic waste products, as satisfactory explanation of their general function has yet to emerge.

ACKNOWLEDGEMENTS

The support of this work by the National Science Foundation, the Washington Mint Commission, the Mint Industry Research Council and Haarmann and Reimer GmbH is gratefully acknowledged.

8. REFERENCES

Adams, R.P., and Hagerman, A., 1977, Diurnal variation in the volatile terpenoids of *Juniperus scopulorum* (Cupressaceae). Am. J. Bot. 64:278-285.

Amelunxen, F., and Arbeiter, H., 1967, Untersuchungen an den Spritzdrüsen von *Dictamnus albus* L. Z. Pflanzenphysiol. 58: 49-69.

Amelunxen, F., Wahlig, T., and Arbeiter, H., 1969, Über den Nachweis des ätherischen Öls in isolierten Drüsenhaaren und Drüsenschuppen von *Mentha piperita* L. Z. Pflanzenphysiol. 61:68-72.

Andersen, N.H., Ohta, Y., and Syrdal, D.D., 1978, Studies in sesquiterpene biogenesis: Implications of absoulate configuration, new structural types, and efficient chemical simulation of pathways. *In:* van Tamelen, E.E., (ed.), Bioorganic Chemistry Vol. II, Substrate Behavior, 1-37, Academic Press, New York.

Arigoni, D., 1975, Stereochemical aspects of sesquiterpene biosynthesis. Pure Appl. Chem. 41:219-245.

Attaway, J.A., Pieringer, A.P., and Barabas, L.J., 1966, The origin of citrus flavor components - I. The analysis of citrus leaf oils using gas-liquid chromatography, thin-layer chromatography and mass spectrometry. Phytochemistry 5:141-151.

Attaway, J.A., Peringer, A.P., and Barabas, L.J., 1967, The origin of citrus flavor components - III. A study of the percentage variations in peel and leaf oil terpenes during one season. Phytochemistry 6:25-32.

Bakker, H.J., Ghisalberti, E.L., and Jefferies, P.R., 1972, Biosynthesis of diterpenes in *Beyeria leschenaultii*. Phytochemistry 11:2221-2231.

Banthrope, D.V., and Charlwood, B.V., 1980, The terpenoids. *In:* Bell, E.A., and Charlwood, B.V. (eds.), Secondary Plant Products. Vol. 8. Encyclopedia of Plant Physiology, 185-220, Springer Verlag, Berlin.

Banthrope, D.V., Charlwood, B.V., and Francis, M.J.O., 1972a, The biosynthesis of monoterpenes. Chem. Rev. 72:115-155.

Banthrope, D.V., and Ekundayo, O., 1976, Biosynthesis of (+)-car-3-ene in *Pinus* species. Phytochemistry 15:109-112.

Banthrope, D.V., Ekundayo, O., Mann J., and Turnbull, K.W., 1975, Biosynthesis of monoterpenes in plants from ^{14}C-labelled acetate and CO_2. Phytochemistry 14:707-715.

Banthrope, D.V., and Whittaker, D., 1966, The preparation and stereochemistry of pinane derivatives. Chem. Rev. 66:643-656.

Banthrope, D.V., and Wirz-Justice, A., 1969, Terpene biosynthesis - Part I. Preliminary tracer studies on terpenoids and chlorophyll of *Tanasetum vulgare* L. J. Chem. Soc. (C): 541-549.

Battaile, J., and Loomis, W.D., 1961, Biosynthesis of terpenes -- II. The site and sequence of terpene formation in peppermint. Biochem. Biophys. Acta 51:545-552.

Battaile, J., Burbott, A.J., and Loomis, W.D., 1968, Monoterpene interconversions: Metabolism of pulegone by a cell-free system from *Mentha piperita*. Phytochemistry 7:1159-1163.

Bellesia., F., Grandi, R., Marchesini, A., Pagnoni, V., and Trave, R., 1975, Biosynthesis of the sesquiterpenoid ageratriol. Phytochemistry 14:1737-1740.

Bennett, R.D., Lieber, E.R., and Heftmann, E., 1967, Biosynthesis of steviol from (-)-kaurene. Phytochemistry 6:1107-1110.

Bernard-Dagan, C., Carde, J.P., and Gleizes, M., 1979, Etude des composés terpéniques au cours de la croissance des aiguilles du pin maritime: Comparaison de données biochemiques et ultrastructurales. Can. J. Bot. 57:255-263.

Bevan, C., Ekong, D., and Okogun, J., 1968, West African timbers. 22. Diterpenes of *Oxystigma oxyphyllum* Harms. J. Chem. Soc. (C):1067-1070.

Beytía, E.D., and Porter, J.W., 1976, Biochemistry of polyisoprenoid biosynthesis. Ann. Rev. Biochem. 45:113-142.

Biollaz, M., and Arigoni, D., 1969, Biosynthesis of coriamyrtin and tutin. J. Chem. Soc. (Chem. Commun.), 633-634.

Birch, A.J., Brown, W.V., Corrie, J.E.T., and Moore, B.P., 1972, Neocembrene-A, a termite trail pheromone. J. Chem. Soc. (Perkin I), 2653-2658.

Bobbitt, J.M., and Segebarth, K.P., 1969, The iridoid glycosides and similar substances. *In:* Taylor, W.I. and Battersby, A.R. (eds.), Cyclopenatoid Terpene Derivatives, 1-146, Marcel Dekker, New York.

Breccia, A., and Badiello, R., 1967, The terpene marrubiin. Z. Naturforschg. 22b:44-49.

Briggs, L.H., and White, G.W., 1975, Constituents of the essential oil of *Araucaria araucana*. Tetrahedrun 31:1311-1314.

Brooks, C.J.W., and Keates, R.A.B., 1972, Biosynthesis of petasin in *Petasites hybridus*. Phytochemistry 11:3235-3245.

Bunton, C.A., and Cori, O., 1978, Mechanistic aspects of terpenoid biochemistry: A cooperative basic research program. Interciencia 3:291-297.

Burbott, A.J., and Loomis, W.D., 1967, Effects of light and temperature on the monoterpenes of peppermint. Plant Physiol. 42:20-28.

Burbott, A.J., and Loomis, W.D., 1969, Evidence for metabolic turnover of monoterpenes in peppermint. Plant Physiol. 44:173-179.

Burbott, A.J., and Loomis, W.D., 1980, Monoterpene interconvesions by cell-free enzymes from mint. Phytochemistry, in press.

Burmeister, J., and von Guttenberg, H., 1960, Die ätherishen Öle als Produkt eines partiell anaeroben Stoffwechsels. Planta Med. 8:1-33.

Cane, D.E., 1980a, The stereochemistry of allylic pryophosphate metabolism. Tetrahedron 36:1109-1159.

Cane, D.E., 1980b, Biosynthesis of sesquiterpenes. *In:* Porter, J.W. (ed.), Polyisoprenoid Biosynthesis, in press, Wiley-Interscience, New York.

Charlwood, B.V., and Banthrope, D.V., 1978, The biosynthesis of monoterpenes. Prog. Phytochem. 5:65-125.

Chayet, L., Rojas, C., Cardemil, E., Jabalquinto, A.M., Vicuña, R., and Cori, O., 1977, Biosynthesis of monoterpene hydrocarbons from [1-^{3}H]neryl pyrophosphate and [1-^{3}H]geranyl pyrophosphate by soluble enzymes from *Citrus limonum*. Arch. Biochem. Biophys. 180:318-327.

Corbella, A., Cariboldi, P., Jommi, G., and Scolastico, C., 1969, Biosynthesis of tutin. J. Chem. Soc. (Chem. Commun.), 634.

Cordell, G.A., 1976, Biosynthesis of sesquiterpenes. Chem. Rev. 76:425-460.

Croteau, R., 1977, Site of monoterpene biosynthesis in *Majorana hortensis* leaves. Plant Physiol. 59:519-520.

Croteau, R., 1980a, Biosynthesis of monoterpenes. *In:* Porter, J.W. (ed.), Polyisoprenoid Biosynthesis, in press, Wiley-Interscience, New York.

Croteau, R., 1980b, The biosynthesis of terpene compounds. *In:* Croteau, R. (ed.), Fragrance and Flavor Substances, in press, D&PS Verlag, West Germany.

Croteau, R., Burbott, A.J., and Loomis, W.D., 1972a, Biosynthesis of mono- and sesquiterpenes in peppermint from glucose-^{14}C and ^{14}CO$_2$. Phytochemistry 11:2459-2467.

Croteau, R., Burbott, A.J., and Loomis, W.D., 1972b, Apparent energy deficiency in mono- and sesquiterpene biosynthesis in peppermint. Phytochemistry 11:2937-2948.

Croteau, R., and Felton, M., 1980a, Substrate specificity of monoterpenol dehydrogenases from *Foeniculum vulgare* and *Tanacetum vulgare*. Phytochemistry 19:1343-1347.

Croteau, R., and Felton, M., 1980b, Conversion of [1-^{3}H$_2$,G-^{14}C]geranyl pyrophosphate to cyclic monoterpenes without loss of tritium. Arch. Biochem. Biophys., in press.

Croteau, R., Felton, M., and Ronald, R.C., 1980a, Biosynthesis of monoterpenes: Conversion of the acyclic precursors geranyl pyrophosphate and neryl pyrophosphate to the rearranged monoterpenes fenchol and fenchone by a soluble enzyme preparation from fennel *(Foeniculum vulgare)*. Arch. Biochem. Biophys. 200:524-533.

Croteau, R. Felton, M., and Ronald, R.C., 1980b, Biosynthesis of monoterpenes: Preliminary characterization of *l-endo*-fenchol synthetase from fennel *(Foeniculum vulgare)* and evidence that no free intermediate is involved in the cyclization of geranyl pyrophosphate to the rearranged product. Arch. Biochem. Biophys. 200:534-546.

Croteau, R., Felton, M., Karp, F., and Kjonaas, R., 1980c, Relationship of camphor biosynthesis to leaf development in sage *(Salvia officinalis)*. Plant Physiol, in press.

Croteau, R., and Hooper, C.L., 1978, Metabolism of monoterpenes: Acetylation of (-)-menthol by a soluble enzyme preparation from peppermint. Plant Physiol. 61:737-742.

Croteau, R., Hooper, C.L., and Felton, M., 1978, Biosynthesis of monoterpenes: Partial purification and characterization of a bicyclic monoterpenol dehydrogenase from sage *(Salvia officinalis)*. Arch. Biochem. Biophys. 188:182-193.

Croteau, R., and Karp, F., 1976a, Biosynthesis of monoterpenes: Enzymatic conversion of neryl pyrophosphate to 1,8-cineole, α-terpineol, and cyclic monoterpene hydrocarbons by a cell-free preparation from sage *(Salvia officinalis)*. Arch. Biochem. Biophys. Res. 176:734-746.

Croteau, R., and Karp, F., 1976b, Enzymatic synthesis of camphor from neryl pyrophosphate by a soluble preparation from sage (*Salvia officinalis*). Biochem. Biophys. Res. Commun. 72: 440-447.

Croteau, R., and Karp, F., 1977a, Biosynthesis of monoterpenes: Partial purification and characterization of 1,8-cineole synthetase from *Salvia officinalis*. Arch. Biochem. Biophys. 179:257-265.

Croteau, R., and Karp, F., 1977b, Demonstration of a cyclic pyrophosphate intermediate in the enzymatic conversion of neryl pyrophosphate to borneol. Arch. Biochem, Biophys. 184:77-86.

Croteau, R., and Karp, F., 1979a, Biosynthesis of monoterpenes: Preliminary characteriztion of bornyl pyrophosphate synthetase from sage (*Salvia officinalis*) and demonstration that geranyl pyrophosphate is the preferred substrate for cyclization. Arch. Biochem. Biophys. 198:512-522.

Croteau, R., and Karp, F., 1979b, Biosynthesis of monoterpenes: Hydrolysis of bornyl pyrophosphate, an essential step in camphor biosynthesis, and hyrolysis of geranyl pyrophosphate, the acyclic precursor of camphor, by enzymes from sage (*Salvia officinalis*). Arch. Biochem. Biophys. 198:523-532.

Croteau, R., and Loomis, W.D., 1972, Biosynthesis of mono- and sesquiterpenes in peppermint from mevalonate-2-^{14}C. Phytochemistry 11:1055-1066.

Croteau, R., and Loomis, W.D., 1975, Biosynthesis and metabolism of monoterpenes. Internatl. Flavours, 292-296.

Croteau, R., and Martinkus, C., 1979, Metabolism of monoterpenes: Demonstration of (+)-neomenthyl-β-D-glucoside as a major metabolite of (-) -menthone in peppermint (*Mentha piperita*). Plant Physiol. 64:169-175.

Darnley-Gibbs, R., 1974, Chemotaxonomy of Flowering Plants. McGill-Queens Univ. Press, Montreal.

Dauben, W.G., Thiessen, W.E., and Resnick, P.R., 1965, Cembrene, a fourteen-membered ring diterpene hydrocarbon. J. Org. Chem. 30:1693-1698.

Dell, B., and McComb, A.J., 1974, Resin production and glandular haris in *Beyeria viscosa* (Labill.) (Euphorbiaceae). Aust. J. Bot. 22:195-210.

Dell, B., and McComb, A.J., 1975, Glandular hairs, resin production and habitat of *Newcastelia viscida* E. Pritzel (Dicrastylidaceae). Aust. J. Bot. 23:373-390.

Dell, B., and McComb, A.J., 1977, Glandular hair formation and resin secretion in *Eremophila* F. Meull (Myoporaceae). Protoplasma, 92:71-86.

Dell, B., and McComb, A.J., 1978, Biosynthesis of resin terpenes in leaves and glandular hairs of *Newcastelia viscida*. J. Exptl. Bot. 28:89-95.

Devon, T.K., and Scott, A.I., 1972, Handbook of Naturally Occurring Compounds. Vol. II. Terpenes, Academic Press, New York-London.

Dro, A.S., and Hefendehl, F.W., 1973, Biogenese des Ätherischeu Öls Von *Ocimum gratissimum*. Planta Med., 24:353-366.

Dueber, M.T., Adolf. W., and West, C.A., 1978, Biosynthesis of the diterpene phytoalexin casbene. Partial purification and characterization of casbene synthetase from *Ricinis communis*. Plant Physiol. 62:598-603.

Duncan, J., 1980, Ph.D. Thesis, UCLA.

Erdtman, M.T., Norin, T., Sumimoto, M., and Morrison, A., 1964, Verticillol, a novel type of conifer diterpene. Tetrahedron Lett. 3879-3886.

Eschemoser, A., Felix, D., Gut, M., Meier, J., and Stadler, P.A., 1959, Some aspects of acid-catalyzed cyclizations of terpenoid polyenes. *In:* Woltstenhome, G.E.W. and O'Conner, M. (eds.), Ciba Foundation Symposium on the Biosynthesis of Terpenes and Sterols, 217-230, Little, Brown, Boston.

Fahn, A., 1979, Secretory Tissues in Plants, 158-222. Academic Press, London-New York.

Francis, M.J.O., 1971, Monoterpene biosynthesis. *In:* Goodwin, T.W. (ed.), Aspects of Terpenoid Chemistry and Biochemistry, 29-51, Academic Press, New York.

Francis, M.J.O., and O'Connell, M., 1969, The incorporation of mevalonic acid into rose petal monoterpenes. Phytochemitry 8:1705-1708.

Frost, R.G., and West, C.A., 1977, Properties of kaurene synthetase from *Marah macrocarpus*. Plant Physiol. 59: 22-29.

Fujita, T., Masuda, I., Takao, S., and Fujita, E., 1976, Biosynthesis of natural products. Part 1. Incorporations of *ent*-kaur-16-ene and *ent*-kaur-16-en-15-one into enmein and oridonin. J. Chem. Soc. (Perkin Trans. I), 2098-2102.

Geissman, T.A., 1973, The biogenesis of sesquiterpene lactones of the compositae. Rec. Adv. Phytochem. 6:65-96.

Gleizes, M., 1976, Biologie des Terpénes Végétaux: Étude des Composés Monoterpéniques et Sesquiterpéniques. Annee. Biol. 15:101-127.

Gleizes, M., 1978, Biosynthése des carbures terpéniques du pin maritime: Essai de localization. C. R. Acad. Sci. (Paris) 286D:543-546.

Goodwin, T.W., 1977, The prenyl lipids of the membranes of higher plants. *In:* Tevini, M. and Lichtenthaler, H.K. (eds.), Lipids and Lipid Polymers in Higher Plants, 29-47, Springer-Verlag, Berlin.

Graebe, J.E., and Hedden, P., 1974, Biosynthesis of gibberellins in a cell-free system. *In:* Schreiber, K., Shütte, H.R., and Sembdner, G. (eds.), Biochemistry and Chemistry of Plant Growth Regulators, 1-16, Acad. Sci. GDR, Inst. Plant Biochem., Halle.

Graebe, J.E., and Ropers, H.J., 1978, Gibberellins. *In:* Letham, D.S., Goodwin, P.B., and Higgins, T.J.V. (eds.), Phytohormones and Related Compounds: A Comprehensive Treatise, Vol. 1: 107-204, Elsevier, Amsterdam.

Hanson, J.R., 1971, The biosynthesis of diterpenes. *In:* Herz, W., Grisebach, H., and Kirby, G.S. (eds.), Progress in the Chemistry of Organic Natural Products, Vol. 29:395-416, Springer-Verlag, New York.

Hanson, J.R., 1972, The di- and sesquiterpenes. Part 1: The diterpenes. *In:* Newman, A.A. (ed.), Chemistry of Terpenes and Terpenoids, 155-199, Academic Press, London-New York.

Hanson, J.R., and White, A.F., 1968, Studies in terpenoid biosynthesis - II. The biosynthesis of steviol. Phytochemistry 7:595-597.

Hedden, P., MacMillan, J., and Phinney, B.O., 1978, The metabolism of the gibberellins. Ann. Rev. Plant Physiol. 29: 149-192.

Hefendehl, F.W., Underhill, E.W., and von Rudloff, E., 1967, The biosynthesis of the oxygenated monoterpenes in mint. Phytochemistry 6:823-835.

Heinrich, G., 1977, Die Feinstruktur und das ätherische Öl eines Drüsenhaares von *Monarda Fistulosa*. Biochem. Physiol. Pflanzen 171:17-24.

Heinrich, G., 1979, Zur Cytologie und Physiologie ätherische Öle erzeugender pflanzlicher Drüsenzellen. *In:* Kubeczka, K.H. (ed.), Vorkommen und Analytik ätherischer Öle, 41-50, Georg Thieme Verlag, Stuttgart.

Henderson, W., Hart, J.W., How, P., and Judge, J., 1970, Chemical and morphological studies on sites of sesquiterpene accumulation in *Pogostemon cablin* (Patchouli). Phytochemistry 9:1219-1228.

Hendrickson, J.B., 1959, Stereochemical implications in sesquiterpene biogenesis. Tetrahedron 7:82-89.

Herout, V., 1971, Biochemistry of sesquiterpenes. *In:* Goodwin, T.W. (ed.), Aspects of Terpenoid Chemistry and Biochemistry, 53-94, Academic Press, London-New York.

Herz, W., 1977, Sesquiterpene lactones in the compositae. *In:* Heywood, V.H., Harborne, J.B., and Turner, B.L. (eds.), The Biology and Chemistry of the Compositae, 337-357, Academic Press, London-New York.

Horster, H., 1979, Monoterpenglykoside, eine Diskussion über ihre biologische Bedeutung und Möglichkeiten zur Synthese dieser Verbindungen. *In:* Kubeczka, K.H. (ed.), Vorkommen und Analytik ätherischer Öle, 34-40, Georg Thieme Verlag, Stuttgart.

Inouye, H., 1978, Neuere Ergebnisse über die Biosynthese der Glucoside der Iridoidreihe. Planta Med., 33:193-216.

Jedlicki, E., Jacob, G., Faini, F., Cori, O., and Bunton, C.A., 1972, Stereospecificity of isopentenyl phosphatase isomerase and prenyl transferase from *Pinus* and *Citrus*. Arch. Biochem. Biophys. 152:590-596.

Karrer, W., 1958, Konstitution und Vorkommen der organischen Pflanzenstoffe (exclusive Alkaloide), Birkhauser Verlag, Basel.

Kúc, J., and Lister, N., 1978, Terpenoids and their role in wounded and infected plant storage tissue. *In:* Kahl, G. (ed.), Biochemistry of Wounded Plant Tissues, 203-242, Walter DeGruyter, New York.

Kullenberg, B., and Bergström, G., 1975, Chemical communication between living organisms. Endeavour 34:59-66.

Lawrence, B.M., Hogg, J.W., and Terhune, S.J., 1971, Les huiles essentielles et leurs constituents. Parfums Cosmet. Savons France 1:256-259.

Lemli, J.A.J.M., 1957, The occurrence of menthofuran in oil of peppermint. J. Pharm. Pharmacol. 9:113-117.

Loomis, W.D., 1967, Biosynthesis and metabolism of monoterpenes. *In:* Pridham, J.B.(ed.), Terpenoids in Plants, 59-82, Academic Press, New York.

Loomis, W.D., and Croteau, R., 1973, Biochemistry and physiology of lower terpenoids. Rec. Adv. Phytochem. 6:147-186.

Loomis, W.D., and Croteau, R., 1980, Biochemistry of terpenoids. *In:* Stumpf, P.K., and Conn, E.E. (eds.), The Biochemistry of Plants, Vol. 4, 363-418, Academic Press, New York.

MacMillan, J., 1971, Diterpenes - the gibberellins. *In:* Goodwin, T.W. (ed.), Aspects of Terpenoid Chemistry and Biochemistry, 153-180, Academic Press, New York.

Malingre, T.M., Smith, D., and Batterman, S., 1969, De isolering en gaschromato-grafische analyse van de vluchtige olie int afzourderlijke klierharen van het labiatentype. Pharm. Weekbl. 104:429-435.

Martin, S.S., Langenheim, J.H., and Zavarin, E., 1976, Biosynthesis of sesquiterpenes in *Hymenaea* inferred from their quantitative co-occurrence. Phytochemistry 15: 113-119.

Martinkus-Taylor, C., and Croteau, R., 1980, Metabolism of monoterpenes: Evidence for compartmentation of *l*-menthone metabolism in peppermint *(Mentha piperita)* leaves. Plant Physiol., in press.

Michie, M.J., and Reid, D.M., 1959, Biosynthesis of complex terpenes in the leaf cuticle and trichomes of *Nicotiana tabacum*. Nature (London) 218:578.

Morikawa, K., Hirose, Y., and Nozoe, S., 1971, Biosynthesis of germacrene-C. Tetrahedron Lett., 1131-1132.

Mothes, K., 1973, Pflanze und Tier. Ein Vergleich auf der Ebene des Sekundärstoffwechsels. Sitzungsber. Österr Akad. Wiss. Math. Naturwiss. Kl. Sonderh. Abf. I 181:1-37.

Müller, C.H., 1971, Phytotoxins as plant habitat variables. In: Biochemical Interactions Among Plants, 64-72, Washington D.C., National Academy of Sciences.

Müller, C.H., and Chou, C., 1972, Phytotoxins: An ecological phase of phytochemistry. In: Harborne, J.B. (ed.), Phytochemical Ecology, 201-216, Academic Press, New York.

Nakanishi, K., 1974, Diterpenes. In: Nakanishi, K., Goto, T., Itô, S., Natori, S., and Nozoe, S. (eds.), Natural Products Chemistry. Vol. 1., 185-312, Academic Press, New York.

Nes, W.R., and McKean, M.L., 1977, Biochemistry of Steroids and Other Isopentenoids, Univ. Park Press, Baltimore.

Nicholas, H.J., 1962a, Biosynthesis of sclareol, β-sitosterol and oleanolic acid from mevalonic acid-2-^{14}C. J. Biol. Chem. 237:1481-1484.

Nicholas, H.J., 1962b, Biosynthesis of β-sitosterol and certain terpenoid substances in *Ocimum basilcium* from melavonic acid-2-^{14}C. J. Biol. Chem. 237:1485-1488.

Nicholas, H.J. 1964, Biosynthesis and metabolism of [^{14}C] sclareol. Biochem. Biophys. Acta 84:80-90.

Oehlschlager, A.C., and Ourisson, G., 1967, A comparison of in vivo and in vitro skeletal transformations of diterpenes. In: Pridham, J.B. (ed.), Terpenoids in Plants, 83-109, Academic Press, New York.

Overton, K.H., and Picken, D.J., 1976, Biosynthesis of bisabolene by callus cultures of *Andrographis paniculata*. J. Chem. Soc. (Chem. Commun.), 105-106.

Paech, K., 1950, Biologie und Physiologie der Sekundaren Pflanzenstoffe, Springer-Verlag, Berlin.

Paknikar, S.K., Bhatwadekar, S.V., and Chakravarti, K.K., 1975, Biogenetically significant components of vetiver oil: Occurrence of (-) - α-funebrene and related compounds. Tetrehedron Letts., 2973-2976.

Parker, W., Roberts, J.S., and Ramage, R., 1967, Sesquiterpene biogenesis. Quart. Rev. Chem. Soc. (London) 21:331-363.

Peréz, L.M., Cori, O., Chayet, L., Rojas, C., and Portilla, G., 1980, Product stereospecificity of prenylsynthetases from orange flavedo (Abst.) Plant Physiol. 65:(suppl), 98:537.

Plouvier, V., 1966, Énantiomrophisme et racémisation chez quelques constituants vegetaux. Phytochemistry 5:955-967.

Potty, V.H. and Bruemmer, J.H., 1970a, Oxidation of geraniol by an enzyme system from orange. Phytochemistry 9:1001-1007.

Potty, V.H. and Bruemmer, J.H., 1970b, Limonene reductase system in the orange. Phytochemistry 9:2319-2321.

Poulose, A.J. and Croteau, R., 1978a, Biosynthesis of aromatic monoterpenes: Conversion of γ-terpinene to p-cymene and thymol in *Thymus vulgaris* L. Arch. Biochem. Biophys. 187:307-314.

Poulose, A.J. and Croteau, R., 1978b, γ-Terpinene synthetase: A key enzyme in the biosynthesis of aromatic monoterpenes. Arch. Biochem, Biophys. 191:400-411.

Poulter, D.C., and Rilling, H.C., 1978, The prenyl transferase reaction. Enzymatic and mechanistic studies of the 1-4 coupling reactions in the terpene biosynthetic pathway. Accts. Chem. Res. 11:307-313.

Poulter, C.D., and Rilling, H.C., 1980, Prenuyltransferases and isomerase. In: Porter, J.W. (ed.), Polyisoprenoid Biosynthesis, in press, Wiley-Interscience, New York.

Qureshi, N., and Porter, J.W., 1980, Conversion of acetyl CoA to isopentenyl pyrophosphate. In: Porter, J.W. (ed.), Polyisoprenoid Biosynthesis, in press, Wiley-Interscience, New York.

Rappaport, L., and Adams, D., 1978, Gibberellins: Synthesis, compartmentation and physiological process, Phil. Trans. R. Soc. Lond. B 284:521-539.

Reid, W.W., 1979, The diterpenes of *Nicotiana* species and *N. tabacum* cultivars. In: Hawkes, J.G., Lester, R.N., and Skelding, A.D. (eds.), The Biology and Taxonomy of the Solanaceae, 273-278, Academic Press, New York.

Regnier, F.E., Waller, G.R., Eisenbraun, E.J., and Auda, H., 1968, The biosynthesis of methylcyclopentane monoterpenoids-II. Nepetalactone. Phytochemistry 7:221-230.

Roberts, J.S., 1972, The sesquiterpenes. In: Newman, A.A. (ed.), Chemistry of Terpenes and Terpenoids, 88-154, Academic Press, New York.

Robinson, D.R., and West, C.A., 1970a, Biosynthesis of cyclic diterpenes in extracts from seedling of *Ricinus communis*. L. I. Identification of deterpene hydrocarbons formed from mevalonate, Biochemistry 9:70-79.

Robinson, D.R., and West, C.A., 1970b, Biosynthesis of cyclic diterpenes in extracts from seedlings of *Ricinus communis* L. II. Conversion of geranylgeranyl pyrophosphate into diterpene hydrocarbons and partial purification of the cyclization enzymes. Biochemistry 9:80-89.

Rogers, L.J., Shah, S.P.J., and Goodwin, T.W., 1968, Compartmentation of biosynthesis of terpenoids in green plants. Photosynthetica 2:184-207.

Rücker, G., 1973, Sesquiterpenes. Angew. Chem. Internat. Ed. 12:793-805.
Ruddat, M. Heftmann, E., and Lang, A., 1965, Biosynthesis of steviol, Arch. Biochem. Biophys. 110:496-499.
Ruzicka, L., 1959, History of the isoprene rule, Proc. Chem. Soc. (London), 341-360.
Ruzicka, L., 1963, Perspektiven der Biogenese und der Chemie der Terpene, Pure and Appl. Chem. 6:493-523.
Ruzicka, L., Eschenmoser, A., and Heusser, H., 1953, The isoprene rule and the biogenesis of terpenic compounds, Experientia 9:357-396.
Sagami, H., Ogura, K., Seto, S., and Kurokawa, T., 1978, A new prenyltransferase from *Micrococcus lysodiekticus*. Biochem. Biophys. Res. Commun. 85:572-578.
Sandermann, W., 1962, Terpenoids: structure and distribution. *In:* Florkin, M., and Mason, H.S. (eds.), Comparative Biochemistry, Vol. 3, Part A, 503-590, Academic Press, New York.
Schnepf, E., 1974, Gland cells. *In:* Robards, A.W. (ed.), Dynamic Aspects of Plant Ultrastructure, 331-357, McGraw-Hill, New York.
Schröder, V., 1969, Untersuchungen zur Tagesrhythmik ätherischer Öle, Pharmazie 24:179, 421.
Scora, R.W., and Mann, J.D., 1967, Essential oil synthesis in *Monarda punctata*. Lloydia 30:236-241.
Scott, A.I., McCapra, F., Comer, F., Sutherland, S.A., Young, D.W., Sim., G.W., and Ferguson, E., 1964, Stereochemistry of the diterpenoids - IV. Structure and stereochemistry of some polycyclic diterpenoids, Tetrahedron 20:1339-1358.
Shah, D.H., Cleland, W.W., and Porter, J.W., 1965, The partial purification, properties, and mechanisms of action of pig liver isopentenyl pyrophosphate isomerase. J. Biol. Chem. 240:1946-1956.
Shechter, I., and West, C.A., 1969, Biosynthesis of of gibberellins - IV. Biosynthesis of cyclic diterpenes from *trans*-geranylgeranyl pyrophosphate. J. Biol. Chem. 244:3200-3209.
Simcox, P.D., 1976, The synthesis of kaurene and related diterpenes in higher plants, Ph.D. Thesis, UCLA. (Dissert. Abs. (B) 37:2218-2219).
Sitton, D., and West, C.A., 1975, Casbene: An anti-fungal diterpene produced in cell-free extracts of *Ricinus communis* seedlings, Phytochemistry 14:1921-1925.
Skopp, K., Hörster, H., 1976, An Zucker gebundene reguläre Monoterpene, Teil I. Thymol und Carvacrolglykosides in *Thymus vulgaris*. Planta Med. 29:208-215.
Smith, D.H., and Carhart, R.E., 1976, Structural isomerism of mono- and sesquiterpenoid skeletons, Tetrahedron 32: 2513-2519.

Souchek, M., 1962, On terpenes. CXLVIII. Biosynthesis of carotol in *Daucas carota* L. A contribution to configuration of carotol and daucol. Coll. Czech. Chem. Comm. 27: 2929-2933.

Sticher, O., and Flück, H., 1968, Die Zusammensetzung von genuinen extrahierten und destillierten ätherischen Ölen einiger *Mentha*-Arten. Pharm. Acta Helv. 43:411-446.

Stoessl, A., Ward, E.W.B., and Stothers, J.B., 1978, Biosynthesis studies of stress metabolites from potatoes: Incorporation of sodium acetate-$^{13}C_2$ into then sesquiterpenes. Can. J. Chem. 56:645-653.

Suga, T., Shishibori, T., and Morinaka, H., 1980, Preferential participation of linaloyl pyrophosphate rather than neryl pyrophosphate in biosynthesis of cyclic monoterpenoids in higher plants. J. Chem. Soc. (Chem. Commun.), 167-168.

Sutherland, M.D., and Park, R.J., 1967, Sesquiterpenes and their biogenesis in *Myoporum deserti* A. Cunn. *In:* Pridham, J.B. (ed.), Terpenoids in Plants, 147-158, Academic Press, New York.

Swain, T., 1974, Biochemical evolution in plants. *In:* Florkin, M., and Stotz, E.H. (eds.), Comprehensive Biochemistry, Vol. 29, Part A, 125-296, Elsevier Pub. Co., New York.

Tetenyi, P., 1970, Infraspecific Chemical Taxes of Medicinal Plants, Chemical Pub. Co., New York.

Upper, C.D., and West, C.A., 1967, Biosynthesis of gibberellins - II. Enzymic cyclization of geranylgeranyl pyrophosphate to kaurene. J. Biol. Chem. 242:3285-3292.

Valenzuela, P., Beytia, E., Cori, O., and Yudelevich, A., 1966, Phosphorylated intermediates of terpene biosynthesis in *Pinus radiata*. Arch. Biochem. Biophys. 113:536-539.

Verzár-Petri, G., and Then, M., 1975, The study of the localization of volatile oil in the different parts of *Salvia sclarea* L. and *Salvia officinalis* L. by applying 2-^{14}C sodium acetate. Acta Botan. Scient. Hung. 21:189-205.

Wallach, O., 1914, Terpene und Camphor, 2nd Ed. Leipzig, Vit.

Weiss, B., and Flück, H., 1970, Untersuchungen über die Variabilität von Gehalt und Zusammensetzung des ätherischen Öles in Blatt und Krautdragen von *Thymus vulgaris* L. Pharm. Acta Helv. 45:169-183.

Wenkert, E., 1955, Structural and biogenetic relationships in the diterpene series, Chem. Ind. (London), 282-284.

West, C.A., 1980a, Hydroxylases, monooxygenases and cytochrome P-450. *In:* Davies, D.D. (ed.), The Biochemistry of Plants, Vol. 2 (Metabolism and Respiration), in press, Academic Press, New York.

West, C.A., 1980b, Biosynthesis of diterpenes, *In:* Porter, J.W. (ed.), Polyisoprenoid Biosynthesis, in press, Wiley-Interscience, New York.

West, C.A., Dudley, M.W., and Dueber, M.T., 1979, Regulation of terpenoid biosynthesis in higher plants, Rec. Adv. Phytochem. 13:163-198.

White, J.D., and Manchard, P.S., 1978, Recent structural investigations of diterpenes. *In:* van Tamelen, E.E. (ed.), Bioorganic Chemistry. Vol. II Substrate Behavior, 337-360, Academic Press, New York.

Whittaker, D., 1972, The monoterpenes. *In:* Newman, A.A. (ed.), Chemistry of Terpenes and Terpenoids, 11-87, Academic Press, New York.

Wuu, T., and Baisted, D.J., 1973, Non-uniform labelling of geraniol biosynthesized from $^{14}CO_2$ in *Pelargonium graveolens*. Phytochemistry 12:1291-1297.

Yafin, Y., and Shechter, I., 1975, Comparison between biosynthesis of *ent*-kaurene in germinating tomato seeds and cell suspension cultures of tomato and tobacco. Plant Physiol. 56:671-675.

Yoshioka, H., Mabry, T.J., and Timmerman, B.N., 1973, Sesquiterpene Lactones, University of Tokyo Press, Tokyo.

Zabkiewicz, J.A., Keates, T.J., and Brooks, C.J.W., 1969, Incorporation of mevalonolactone into *Petasites hybridus*: Effect of synthetic inhibitors on sesquiterpenoid and sterol production. Phytochemistry 8:2087-2089.

Zavarin, E., 1970, Qualitative and quantitative co-occurrence of terpenoids as a tool for elucidation of their biosynthesis. Phytochemistry 9:1049-1063.

8

THE CHEMISTRY OF BIOLOGICALLY ACTIVE CONSTITUENTS SECRETED AND STORED IN PLANT GLANDULAR TRICHOMES

Rick G. Kelsey

Woods Chemistry Laboratory
Department of Chemistry
University of Montana
Missoula, Montana 59823

and

Gary W. Reynolds and Eloy Rodriguez

Phytochemistry Laboratory
Department of Developmental and Cell Biology
and
Ecology and Evolutionary Biology
University of California, Irvine
Irvine, California 92717

ABSTRACT

A diverse group of small molecular weight chemical constituents constitute the major natural products present in glandular hairs of plant surfaces. Although isoprenoids, which include the monoterpenes, sesquiterpenes and diterpenes, are the most commonly occurring metabolites in trichomes of flowering plants, other secondary metabolites are present but restricted to a limited number of plant families. For example, taxa of **Larrea** *(Zygophyllaceae) contain copious amounts of methylated flavonoids and lignans which function in defense, both against herbivores and the harsh arid environment. Alkylated phenolics and quinones are major trichome products in members of the Hydrophyllaceae and Primulaceae, with many of the compounds found to be allergenic to humans. In this review, the chemistry and distribution of compounds demonstrated to be present in trichomes is described, with special reference to their biological activities and possible ecological functions.*

1. INTRODUCTION

The importance of glandular trichomes to the success of higher plants is obvious when one considers the widespread occurrence of these structures. Of all the species of plants included in Abrams' *Illustrated Flora of the Pacific States* (1940-1960) and Munz's *A California Flora* (1959), between 20 and 30 percent are described as having glandular hairs. However, the specific function of the glandular trichomes of a given species is generally not apparent. In some cases the functions are obvious, such as with the salt glands of halophytes, the trapping glands of carnivorous plants, and the secretory trichomes of nectaries. In other cases the advantage to the plant is not clear, particularly concerning trichomes that apparently play some role in the plant's defense against herbivory or disease. A case for the role of trichomes in plant defense has been put forth by Levin (1973). He discussed the difficulties in specifying defensive roles to trichomes which result from the complex and dynamic nature of each species' interactions with other organisms, and also from the paucity of information regarding the chemistry and biological activity of the constituents of glandular trichomes. In a voluminous and expanding phytochemical literature only infrequently do chemists report the localization of natural products beyond gross aspects of plant anatomy. A wealth of information has accrued from botanists describing the anatomy, development, and ultrastructure of glandular trichomes (Uphof 1962, Schnepf 1969, Lüttge 1971, Fahn 1979), and increasingly from entomologists come reports describing correlations of resistance to insect damage with the presence of glandular trichomes. However, here the great majority of chemical information specific to trichomes has been from histochemical techniques which provide the chemical class, but not the specific identity of the active constituents. In order for theories to be developed which describe the evolution and *raison d'etre* of glandular trichomes in plants, the identity of the gland constituents and their biological activity must be established. The purpose of this chapter is to review the information to date on the chemistry of compounds definitely known to be secreted and stored in trichomes in amounts significant to the potential function of the trichomes.

2. TERPENOIDS

Terpenoids represent one of the largest and most structurally diverse classes of natural products with some type being present in every member of the plant kingdom (Nicholas 1973). Although ubiquitous to plant tissue, many plants have developed specialized structures, glandular or secretory cells, oil or resin canals, and glandular trichomes (Esau 1965) where the terpenes accumulate or are secreted in high concentration. Terpenoids are the most commonly encountered class of natural products not only in glandular trichomes but in all specialized secretory structures.

Some terpenes stored or secreted from glandular trichomes are listed in Tables 1 and 2 and Figures 1 to 5. In compiling this list we used only those plant species studied in sufficient detail to establish that the terpenoids were associated with the trichomes. Not included were those species that are known to have glands and also produce terpenoids (in particular, many species in the family Labiatae), although there is little doubt that these compounds occur in the glands.

Monoterpenes and sesquiterpenes are almost always linked with secretory cells or structures (Loomis and Croteau 1973), and they are the most common terpenoid components of glandular trichomes. It is not unusual to find them mixed in the glandular secretions, but their sites of biosynthesis are not the same and they appear to be compartmentalized, probably within the glandular hairs (Loomis and Croteau 1973, Croteau 1977).

Diterpenes and triterpenes are in the resinous glandular exudates that cover the leaf surfaces of many plants from the semi-arid regions of western Australia. Dell (1977) examined 32 resinous genera from 22 plant families and found only 5 that lacked the glandular hairs, indicating a strong correlation between the resins and trichomes. There is some experimental evidence that the biosynthesis of these terpenoids occurs within the glands (Michie and Reid 1968, Dell and McComb 1978a). Table 1 is by no means a comprehensive list of all the di- and triterpenoids in resinous exudates, but rather centers around the more thoroughly studied genera *Beyeria*, *Eremophilia* and *Newcastelia* (Dell and McComb 1978b).

Tetraterpenes or carotenoids do not accumulate to high concentrations in specialized cells or glandular trichomes, but rather are associated with tissues of flowers, fruits and the chlorophylls in the grana of chloroplasts in green plants (Goodwin 1967, Ramage 1972).

Included in this discussion are the mixed-terpenoids or compounds whose structures are partially derived from isopentenyl pyrophosphate (Goodwin 1967). These glandular products have been isolated from trichomes of hops, *Humulus lupulus* L., and its close relative marijuana, *Cannabis sativa* L., in combination with the monoterpenes and sesquiterpenes (Guenther 1952, Stevens 1967, Hendriks et al. 1975, Malingre et al. 1975, Turner et al. 1980).

2.1 *Insecticides, Repellents and Insect Feeding Deterrents*

Terpenoids are not only important to plants, but they are also very important within the world of insects. The growth and

Table 1. Terpenoids that occur as major* components in glandular trichomes

Compound	Species	Reference
MONOTERPENES		
Acyclic		
myrcene (1)	*Humulus lupulus* L.	Buttery & Ling 1967.
nerol (2)	*Humulus spinosa* L.	Amelunxen & Arbeiter 1969.
Monocyclic		
menthol (3)	*Mentha piperita* L.	Sticher & Fluck 1968.
menthone (4)	*M. piperita*	Sticher & Fluck 1968, Amelunxen et al. 1969.
menthylacetate (5)	*M. piperita*	Sticher & Fluck 1968.
Bicyclic		
camphene (6)	*Artemisia tridentata* Nutt. spp. *vaseyana* (Rydb.) Beetle	Kelsey & Shafizadeh 1980, Scholl 1976.
	A. nova Nelson	
camphor (7)	*A. tridentata* ssp. *vaseyana*	Kelsey & Shafizadeh 1980, Scholl 1976.
	A. nova	
cineole (8)	*A. tridentata* ssp. *vaseyana*	Kelsey & Shafizadeh 1980, Scholl 1976.
	Mentha piperita	Amelunxen et al. 1969.

Table 1 (continued)

Compound	Species	Reference
menthofuran (9)	*M. piperita*	Sticher & Fluck 1968.
	M. aquatica L.	Malingre et al. 1969.
nepetalactone (10) and epinepetalactone (11)	*Nepeta cataria* L.	Regnier et al. 1967, 1968.
α-pinene (12)	*Artemisia tridentata* ssp. *vaseyana*	Kelsey & Shafizadeh 1980, Scholl 1976.
	A. nova	
curcumene (22)	*Cannabis sativa*	Nigam et al. 1965, Hood et al. 1973.
β-farnescene (23)	*C. sativa*	Nigam et al. 1965, Hood et al. 1973, Hendriks et al. 1975.
	Humulus lupulus	Buttery et al. 1967, Buttery & Ling 1967.
gerin (24)	*Geraea viscida* (Gray) Blake	Rodriguez et al. 1979.
α-guaiene (25)	*Pogostemon cablin*	Henderson et al. 1970.
β-humulene (26)	*Cannabis sativa*	Nigam et al. 1965., Hood et al. 1973, Hendriks et al. 1975.
	Humulus lupulus	Buttery et al. 1967, Buttery & Ling 1967.
isocaryophyllene (27)	*Nepeta cataria*	Regnier et al. 1967.
	Cannabis sativa	Hendriks et al. 1975.
longifolene (28)	*C. sativa*	Stahl & Kunde 1973.

Table 1 (continued)

Compound	Species	Reference
α-patchoulene (29)	*Pogostemon cablin*	Henderson et al. 1970.
β-patchoulene (30)	*P. cablin*	Henderson et al. 1970.
γ-patchoulene (31)	*P. cablin*	Henderson et al. 1970.
patchouli alcohol (32)	*P. cablin*	Henderson et al. 1970.
selina-3,7(11)-diene (33)	*Cannabis sativa* *Humulus lupulus*	Hendriks et al. 1975. Buttery et al. 1967, Buttery & Ling 1967.
selina-4(14),7(11)-diene (34)	*Cannabis sativa* *Humulus lupulus*	Hendriks et al. 1975. Buttery et al. 1967, Buttery & Ling 1967.
sabinene (13)	*Majorana hortensis* Moench	Croteau 1977.
cis-sabinene hydrate (14)	*M. hortensis*	Croteau 1977.
thujone (15)	*Artemisia tridentata* spp. *vaseyana*	Kelsey & Shafizadeh 1980, Scholl 1976.
Irregular		
artemeseole (16)	*A. tridentata* ssp. *vaseyana*	Kelsey & Shafizadeh 1980, Scholl 1976, Noble & Epstein 1977.

Table 1 (continued)

Compound	Species	Reference
SESQUITERPENES		
Hydrocarbons		
trans-α-bergamotene (-17)	Cannabis sativa L.	Nigam et al. 1965, Hood et al. 1973, Hendriks et al. 1975.
β-bisabolene (18)	C. sativa	Hood et al. 1973.
bulnesene (19)	Pogostemon cablin Bench.	Henderson et al. 1970.
δ-cadinene (20)	Humulus lupulus	Buttery et al. 1967, Buttery & Ling 1967.
β-caryophyllene (21)	Cannabis sativa	Nigam et al. 1965, Hood et al. 1973, Hendriks et al. 1975.
	Cleome spinosa	Amelunxen & Arbeiter 1969.
	Humulus lupulus	Buttery et al. 1967, Buttery & Ling 1967.
	Mentha aquatica	Malingre et al. 1969.
	Nepeta cataria	Regnier et al. 1967.
α-selinene (35)	Cannabis sativa	Nigam et al. 1965, Hood et al. 1973.
	Humulus lupulus	Buttery et al. 1967, Buttery & Ling 1967.

Table 1 (continued)

Compound	Species	Reference
β-selinene (36)	*Cannabis sativa*	Hendriks et al. 1975.
	Humulus lupulus	Buttery et al. 1967, Buttery & Ling 1967.
Lactones		
absinthin (37)	*Artemisia absinthium* L.	Politis 1946, Kelsey & Shafizadeh 1979.
ambrosin (38)	*Parthenium hysterophorus* L.	Rodriguez et al. 1976a.
anabsinthin (39)	*Artemisia absinthium*	Politis 1946, Kelsey & Shafizadeh 1979.
artabsin (40)	*A. absinthium*	Politis 1946, Kelsey & Shafizadeh 1979.
cnicin (41)	*Cnicus benedictus* L.	Politis 1946, Vanhaelen-Fastre 1972.
cumambrin-A (42)	*Artemisia nova*	Kelsey & Shafizadeh 1980.
cumambrin-B (43)	*A. nova*	Kelsey & Shafizadeh 1980.
parthenin (44)	*Parthenium hysterophorus*	Rodriguez et al. 1976a.
parthenolide (45)	*Chrysanthemum parthenium* Bernh.	Blakeman & Atkinson 1979.

Table 1 (continued)

Compound	Species	Reference
DITERPENES		
Monocyclic		
cembrenetriol (46)	*Eremophila clarkei* F. Muell.	Coates et al. 1977.
an -4,8,13-duvatriene-1,3-diol (47)	*Nicotiana tabacum* L.	Roberts & Rowland 1962, Michie & Reid 1968.
epoxycembradienol (48)	*Eremophila georgei* Diels	Ghisalberti et al. 1977.
Bicyclic		
eperuane-8,15-diol (49)	*Ricinocarpos muricatus* F. Muell	Dell & McComb 1978b.
dihydroxyserrulatic acid (50)	*Eremophila serrulata* (F. Muell.) Druce	Croft et al. 1977.
Tricyclic		
beyer-15-en-3-one-17-oic-acid,19-OAc (51)	*Beyeria leschenaultii* Baill.	Bakker et al. 1972.
beyer-15-en-3-one,β OAc-17-ol (52)	*B. leschenaultii*	Bakker et al. 1972.

Table 1 (continued)

Compound	Species	Reference
beyer-15-en-3-one, β 17-diol (53)	B. leschenaultii	Bakker et al. 1972.
decipiane triol (54)	Eremophila decipiens Ostenf.	Dell & McComb 1978b, Ghisalberti et al. 1975.
eremolactone (55)	E. fraseri F. Muell.	Dell 1975, Dell & McComb 1978b, Birch et al. 1966.
isopimara-9(11),15-diene-3 19-diol (56)	Newcastelia viscida E. Pritzel	Jefferies & Ratajczak 1973, Dell & McComb 1975, 1978a,b.
3,4 secobeyerene acid (57)	Beyeria leschenaultii	Bakker et al. 1972.
TRITERPENES		
betulic acid (58)	Newcastelia viscida	Dawson et al. 1966, Dell & McComb 1975.
oleanolic acid (59)	N. viscida	Dawson et al. 1966, Dell & McComb 1975, 1978b.
MIXED-TERPENES		
adhumulone (60)	Humulus lupulus	Stevens 1967.
adlupulone (61)	H. lupulus	Stevens 1967.

Table 1 (continued)

Compound	Species	Reference
cannabichromene (62)	Cannabis sativa	Willinsky 1973, Malingre et al. 1975, Turner et al. 1977, 1978.
cannabidiol (63)	C. sativa	Willinsky 1973, Malingre et al. 1975, Turner et al. 1977, 1978.
cannabigerol (64)	C. sativa	Willinsky 1973, Malingre et al. 1975.
cannabinol (65)	C. sativa	Willinsky 1973, Malingre et al. 1975, Turner et al. 1977, 1978.
cohumulone (66)	Humulus lupulus	Stevens 1967.
colupulone (67)	H. lupulus	Stevens 1967.
humulone (68)	H. lupulus	Guenther 1952, Stevens 1967.
lupulone (69)	H. lupulus	Guenther 1952, Stevens 1967.
Δ^9-tetrahydrocannabinol (70)	Cannabis sativa	Willinsky 1973, Malingre et al. 1975, Turner et al. 1977, 1978.

* The determination of a major component was arbitrary, but as a rule the monoterpenes had to constitute at least 10% of the oil and sesquiterpenes 1%.

Table 2. Terpenoids that occur as minor* components in glandular trichomes

MONOTERPENES

borneol
carvacrol
p-cymol
dihydronepetalactone
geranyl acetate
geranyl isobutyrate
geranyl propionate
isomenthone
limonene
linalool
linalylacetate
linalylpropionate
linalylvalerianate
menthene
menthylacetate
methyl geranate
neomenthol
nerolidiol
ocimene
β-pinine
piperitone

MONOTERPENES (Cont'd.)

pulegone
trans-sabinene hydrate
α-terpinene
γ-terpinene
α-terpineol
α-thujene
thymol

SESQUITERPENES

γ-cadinene
copaene
humulene epoxide
humulenol
β-santalane
selina-4(5),7(11)-diene

DITERPENES

beyer-15-en-19-ol
beyerol
17-hydroxy-19-norbeyer-15-en-3-one

*The determination of a minor component was arbitrary, but as a rule the monoterpenes had to consitute less than 10% of the oil and sesquiterpenes less than 1%. These compounds were reported in one of the following references: Buttery et al. 1967; Buttery & Ling 1967; Regnier 1967, 1968; Sticher & Fluck 1968; Amelunxen et al. 1969; Amelunxen & Arbeiter 1969; Malingre et al. 1969; Bakker et al. 1972; Croteau 1977. For a list of all the natural constituents that have been isolated from marijuana, see Turner et al. 1980.

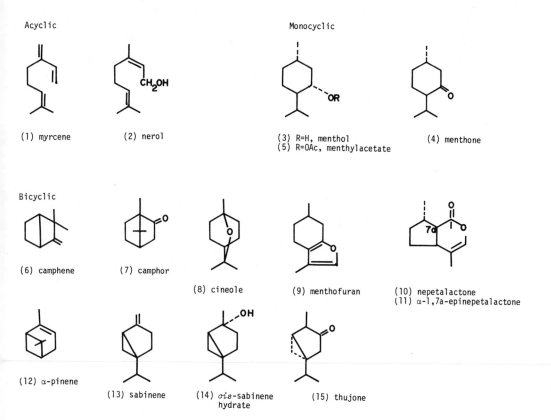

Figure 1. Monoterpenes in glandular trichomes.

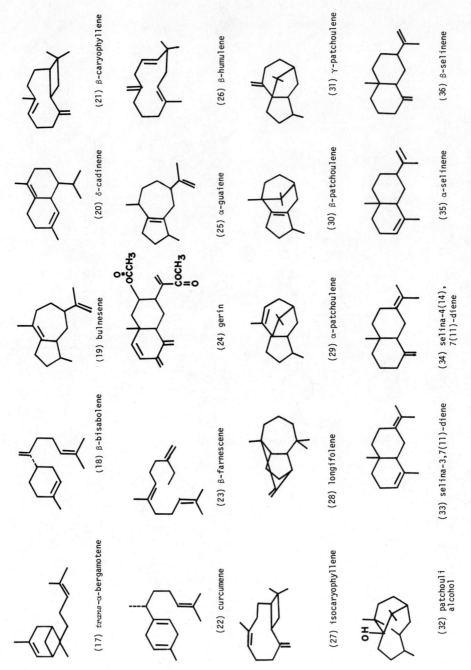

Figure 2. Sesquiterpenes in glandular trichomes.

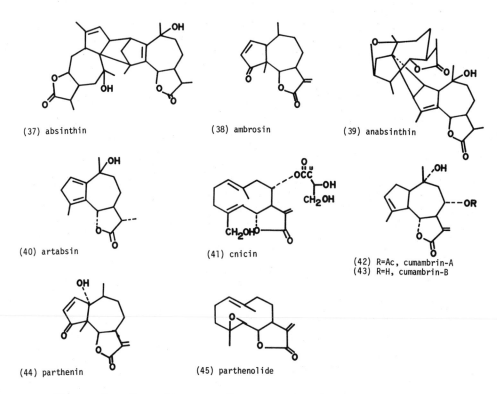

Figure 3. Sesquiterpene lactones in glandular trichomes.

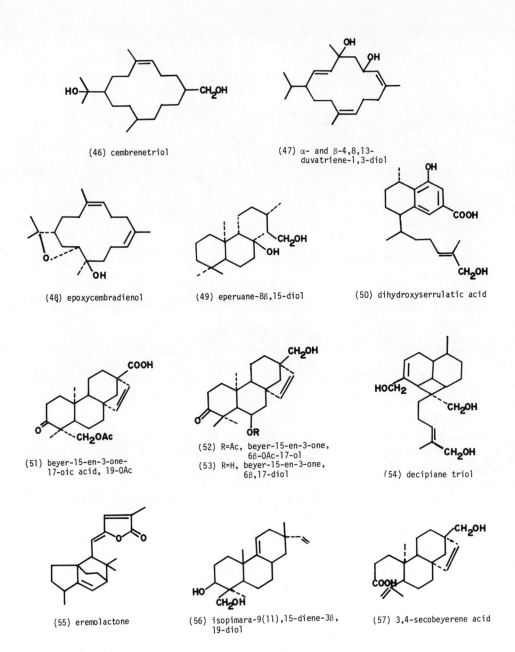

Figure 4. Monocyclic, Bicyclic, and Tricyclic Diterpenes.

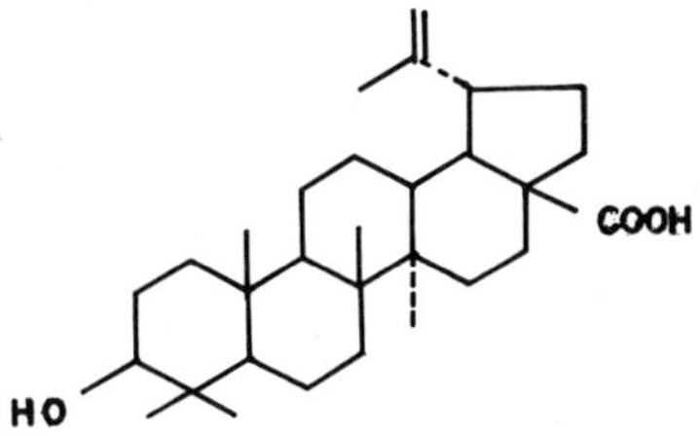

BIOLOGICALLY ACTIVE CONSTITUENTS IN PLANT GLANDULAR TRICHOMES   203

(58) betulic acid

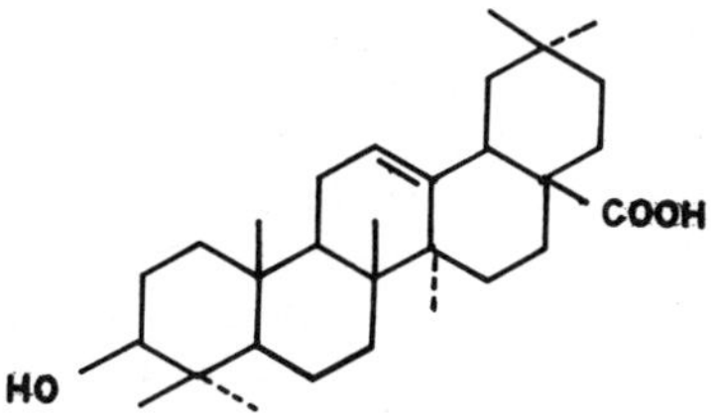

(59) oleanolic acid

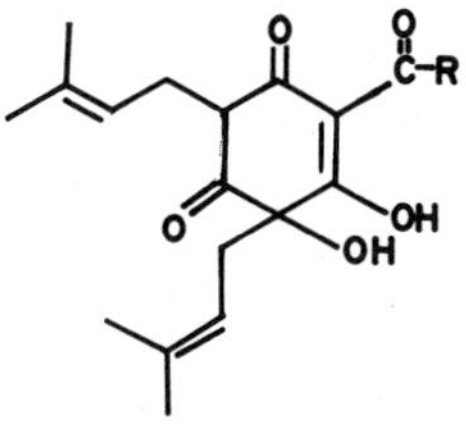

(60) R=sec-butyl, adhumulone
(66) R=i-propyl, cohumulone
(68) R=i-butyl, humulone

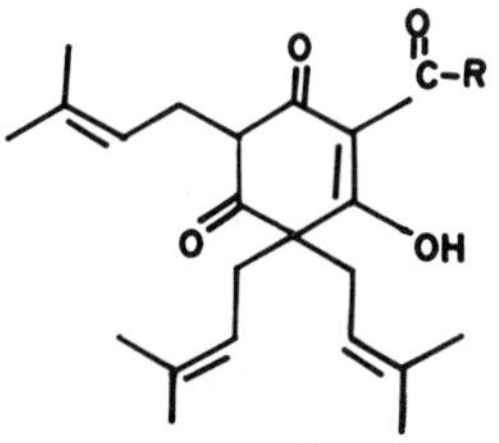

(61) R=sec-butyl, adlupulone
(67) R=i-propyl, colupulone
(69) R=i-butyl, lupulone

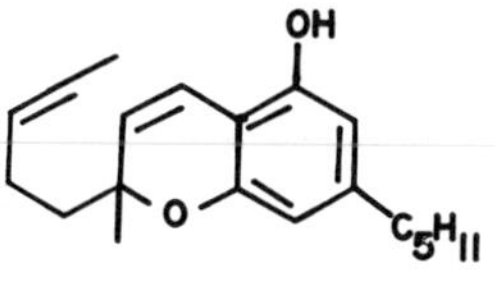

(62) cannabichromene

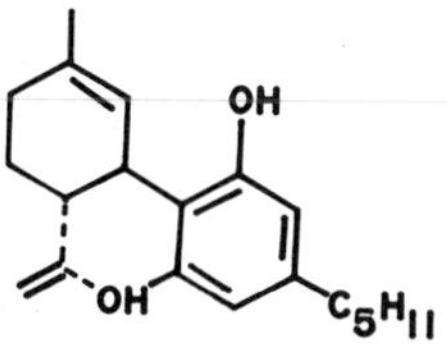

(63) cannabidiol

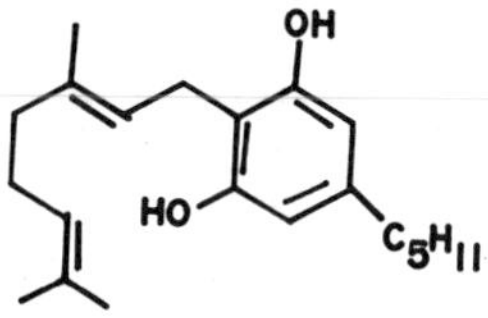

(64) cannabigerol

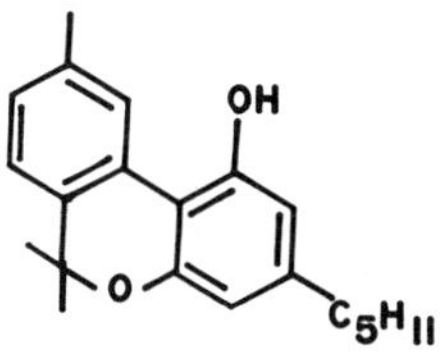

(65) cannabinol

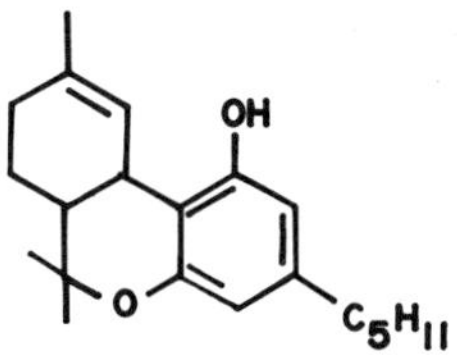

(70) Δ^9-tetrahydrocannabinol

Figure 5. Triterpenes and Mixed Terpenes.

development of insects is regulated by terpenoid hormones, and their behavior can be modified by terpene pheromones (chemical signals) that attract, repel, cause alarm or initiate defensive reactions. Some insects have developed chemical defense systems that utilize these compounds for protection (Weatherston 1967, Karlson 1970, Clayton 1970, Rodriguez and Levin 1976, Harborne 1977). As a result of this chemical parallelism the terpenoid components of a plant can have a significant effect on its ecological association with insects.

The monoterpene α-pinene (12) occurs in the secretions of various insects (Weatherston 1967, Karlson 1970), as well as in the essential oils of angiosperms, and in the oleoresins of gymnosperms. Its effect on insects is variable, ranging from an attractant to that of a toxin, with the specific response dependent upon the insect species, concentration, associated chemicals and mode of contact (Table 3). Myrcene (1) and camphene (6) are similar to α-pinene (12) in their effect on insects, both occurring in the defense secretions of termites (Prestwich 1979) (see Table 4). Camphene (6) is also one of 4 monoterpenes in the alarm pheromone of ants, *Pristomyremex pungens* (Hayashi and Komae 1977). Myrcene (1) is toxic to house flies, *Musca domestica* (Hrdy et al. 1977), while camphene (6) is lethal to the larch sawfly larvae, *Pristiphora erichsonii* (Oshkaev 1977).

The larvae of silkworms, *Bombyx mori*, are repelled by menthol (3) (Ishikawa and Hirao 1966), while the larval development of the European bug, *Pyrrhocoris apterus*, is inhibited by the ketone, menthone (4) (Slama 1978). Both compounds occur in the trichomes of *Mentha piperita* L. as well as other species in the mint family Labiatae (Sticher and Flück 1968, Amelunxen et al. 1969).

The cyclopentanoid monoterpene, nepetalactone (10), the major constituent of catnip, *Nepeta cataria* L. (Regnier et al. 1967, 1968), is an active repellent, producing avoidance responses from insects representing 15 genera in 13 families. The aggressive behavior of ants, *Pogonomyremex badius* was drastically altered by the application of nepetalactone (10) to the body of its prey (Eisner 1964). Other compounds of similar structure, iridoids, have been isolated from alarm-defense secretions of various ant genera (Eisner 1964, Weatherston 1967, Rodriguez and Levin 1976).

Caryophyllene (21), one of the most commonly occurring sesquiterpenes in plants and glandular trichomes (Table 1), inhibits the hormonally controlled development of the European bug, *Pyrrhocoris apterus* (Slama 1978). The major component in the defense secretion of the butterfly larvae, *Battus polydamas*, has been identified as β-selinene (36) (Eisner et al. 1971), one of several sesquiterpenes in the glandular hairs of hops, *Humulus lupulus* (Guenther 1952, Buttery et al. 1967, Buttery and

Table 3. The effects of α-pinene on insects

Insect	Effect	Reference
ANTS		
Pristomyrmex pungens	Alarm	Hayashi & Komae 1977.
Pogonomyrmex occidentalis	Repellent	Eisner et al. 1974.
Formica rufa	Topical toxin	Howse 1975.
Myrmicaria eumenoides	Attractant, alarm, toxin (concentration dependent)	Howse et al. 1977.
BARK BEETLES		
Ips typographus	Repellent	Vasechko 1978.
Hylurgops glabratus	Repellent	Vasechko 1978.
Dendroctonus micans	Attractant	Vasechko 1978.
Ips grandicollis	Attractant	Werner 1972.
FLIES		
Musca domestica	Topical toxin	Hrdy et al. 1977.
SAWFLIES		
Pristiphora erichosonii	Topical and vapor toxin	Oshkaev 1977.
TERMITES		
Nasutitermes rippertii	Alarm	Hrdy et al. 1977, Vrkoc et al. 1978.
N. costalis	Alarm	Hrdy et al. 1977, Vrkoc et al. 1978.

Table 4. Terpenoids from glandular trichomes that affect insects--toxins, repellents and growth regulators.

Compound	Effect	Insects Affected*	Reference
MONOTERPENES			
camphene (6)	alarm	ants, *Pristomymex pungens*	Hayashi & Komae 1977.
	defense secretions	termites, *Trinervitermes bettonianus*	Prestwich 1979.
	toxin	larch sawfly larvae, *Pristiphora erichsonii*	Oshkaev 1977.
menthol (3)	repellent	silkworm larvae, *Bombyx mori*	Ishikawa & Hirao 1966.
menthone (4)	growth inhibitor	European bug, *Pyrrhocoris apterus*	Slama 1978.
myrcene (1)	defense secretions	termites, *Amitermes vitiosus*	Prestwich 1979.
	toxin	house flies, *Musca domestica*	Hrdy et al. 1977.
nepetalactone (10)	repellent	variety of insects	Eisner 1964.
α-pinene (12)	see Table 3		
SESQUITERPENES			
β-caryophyllene (21)	growth inhibitor	European bug, *Pyrrhocoris apterus*	Slama 1978.
β-selinene (36)	defense secretions	butterfly larvae, *Battus polydamas*	Eisner et al. 1971.

*These are examples of insects affected by the compound as described in the "Effect" column except for the defense secretions, in which case the organism listed uses this compound as part of its chemical defense.

Ling 1967) and marijuana, *Cannabis sativa* (Malingre et al. 1975, Hendriks et al. 1975).

The leaf pocket resin of *Hymenaea courbaril* L. is composed of a mixture of the sesquiterpene hydrocarbons caryophyllene (21), α- and β-selinene (35, 36), α- and β-humulene (26), and δ-cadinene (20), among others. This mixture possesses toxic and feeding deterrent properties against the generalist herbivore, beet armyworm, *Spodoptera exigua* (Stubblebine and Langenheim 1977). There are various examples of sesquiterpene lactones being insect feeding deterrents (Burnett et al. 1974, Picman et al. 1978, Doskotch et al. 1980), but none of the compounds in Table 1 from trichomes has as yet been tested.

Soldiers of the termite species *Cubitermes umbratus* have a glandular defense secretion with 4 diterpene hydrocarbons, two of which have been identified as cubitene (71) and neocembrene-A (72) (Figure 6) (Prestwich 1979). The latter compound is also a trail pheromone for a species of Australian termites, *Nasutitermes* (Birch et al. 1972). Diterpenes, structurally similar to the above, cembrenetriol (46) and epoxycembradienol (48), have been isolated from the glandular resins that coat the leaves of *Eremophila clarkei* F. Muell. (Coates et al. 1977) and *E. georgei* Diels (Ghisalberti et al. 1977, Maslen et al.1977), respectively. The effects of these resin diterpenes on insect feeding or other physiological and behavior characteristics are unknown.

The ecological interactions between plants and insects are very complex, with secondary products playing an important role in mediating the interactions. Some volatile terpenoids may act as attractants luring pollinating insects to the flowers for their help in the exchange of genetic material and fertilization, while other compounds with antifeeding properties, such as sesquiterpene lactones, may protect the plant and its reproductive structures from destruction by herbivores (Harborne 1977).

2.2 *Antifungal, Antimicrobial and Cytostatic Activities*

Essential oils, many containing terpenoids, were the first preservatives used by man, originally in their natural state within plant tissues and then as oils obtained by distillation (Cade 1954). Many of these crude mixtures are antifungal and antimicrobial, having been tested with a variety of organisms (Maruzzella and Liguori 1958, Maruzzella and Henry 1958, Maruzzella and Sicurella 1960, Morris et al. 1979). To determine the active components of these oils and plant tissues, individual compounds have been bioassayed after isolation and purification.

Morris et al. (1979) screened 521 fragrance materials and found 44% of them to inhibit at least one of the assay species, *Candida albicans, Escherichia coli, Staphylococcus aureus* and a lipophilic diphtheroid (probably *Corynebacterium* spp.). Among the active materials were th
e three glandular monoterpenes, nerol (2), menthol (3), and camphor (7) and the sesquiterpene, guaiene

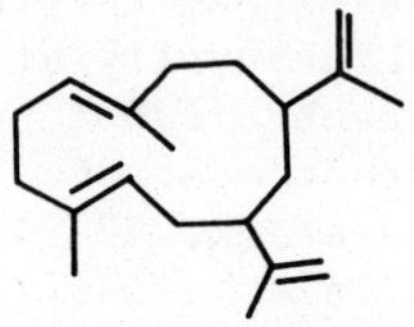

(71) cubitene 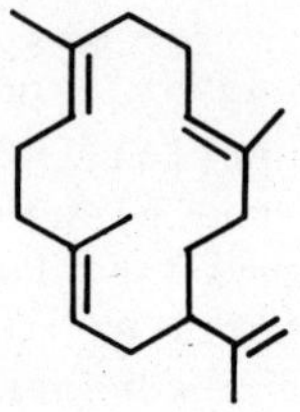 (72) neocembrene-A

Figure 6. Diterpenes isolated from the defense secretion of the termite, *Cubitermes umbratus* (Prestwich, 1979).

(25) (Table 5). Some of the antifungal monoterpenes, myrcene (1), camphene (6), and α-pinene (12) and the sesquiterpene, longifolene (28) occur in oleoresins (Mutton 1962, Buchanan 1963) and retard the growth of fungi associated with trees, including decay organisms, *Fomes annosus, Lenzites saepiaria, Schizophyllum commune* (Cobb et al. 1968, DeGroot 1972, Schuck 1977), ectomycorrhizae symbiotic on the roots of pine, *Boletus variegatus, Rhizopogon roseolus* (Melin and Krupa 1971) and several blue staining fungi in the genus *Ceratocystis* (Cobb et al. 1968, DeGroot 1972). Cobb et al. (1968) reported the volatile terpenoids from *Pinus ponderosa* Laws. to be inhibitory but nontoxic since normal fungal growth resumed upon removal from the vapors. The experimental evidence suggests that the terpenoids in the oleoresin and wood of trees in conjunction with other compounds such as the phenols constitute a chemical defense that protects against fungal infections (Cobb et al. 1968, DeGroot 1972).

The activities of several glandular sesquiterpene lactones have been examined in detail (Table 5). Parthenolide (45) in glandular trichomes of *Chrysanthemum parthenium* Bernh. inhibited the growth of several gram positive bacteria, yeast and filamentous fungi *in vitro*, whereas gram negative bacteria were unaffected (Blakeman and Atkinson 1979). Infection by two pathogenic fungi on chrysanthemum petals and bean leaves was reduced in the presence of this compound (Table 5). Parthenolide (45) is also a skin irritant (Hausen and Schulz 1976) and cytotoxic to several neoplastic tissues, neither of which may be of biological significance to the plant, but these facts do demonstrate the wide ranging activity of such a compound in biological systems. There is a correlation between antimicrobial and cytotoxic activities for sesquiterpene lactones (Lee et al. 1971, Lee et al. 1977). Ambrosin (38) and parthenin (44) are in the trichomes of *Parthenium hysterophorus* L. (Rodriguez et al. 1976a,b), the former being cytotoxic (Torrance et al. 1975) and the latter inhibiting the sporangial germination and zoospore mobility of *Sclerospora graminicola* (Char and Shankarabhat 1975). Both of these lactones have been implicated as allergens causing contact dermatitis in humans (Lonkar et al. 1974, Rodriguez et al. 1976a, b, 1977). Cnicin (41) also appears to be quite active (Vanhaelen-Fastre 1972, Vanhaelen-Fastre and Vanhaelen 1976, Gonzalez et al. 1978).

The only triterpene in Table 5 is oleanolic acid (59), a component of the resinous exudate of *Newcastelia viscida* E. Pritzel, whose glycoside is inhibitory to the yeast *Saccharomyces carlsbergensis* (Anisimov et al. 1979). Many of the mixed-terpenoids in the glandular hairs of *Humulus lupulus* and *Cannabis sativa* retard the growth of gram positive bacteria (Hansens 1947, Hough et al. 1957, Mechoulam and Gaoni 1965, Mechoulam and Edery 1973).

Table 5. Antimicrobial, antifungal and cytostatic terpenoids in glandular trichomes

Compound	Effect	Affected Organism	Reference
MONOTERPENES			
camphene (6)	antifungal	*Fomes annosus* *Lenzites saepiaria* *Schizophyllum commune* *Trichoderma viride*	Cobb et al. 1968, DeGroot 1972.
camphor (7)	antimicrobial	*Staphylococcus aureus*	Morris et al. 1979.
menthol (3)	antifungal antimicrobial	*Candida albicans* *Escherichia coli*	Morris et al. 1979. Morris et al. 1979.
myrcene (1)	antifungal	*Fomes annosus* *Ceratocystis minor* *Lenzites saepiaria* *Schizophyllum commune* *Trichoderma viride*	Cobb et al. 1968, DeGroot 1972.
nerol (2)	antifungal antimicrobial	*Candida albicans* *Escherichia coli* *Staphylococcus aureus* a lipophilic diphtheroid	Morris et al. 1979. Morris et al. 1979

Table 5 (continued)

Compound	Effect	Affect Organism	Reference
α-pinene (12)	antifungal	Boletus variegatus Ceratocystis monor Fomes annosus Lenzites saepiaria Rhizopogon roseolus Schizophyllum commune Trichoderma viride	Cobb et al. 1968, Melin & Krupa 1971, DeGroot 1972, Schuck 1977.
SESQUITERPENES			
ambrosin (38)	Cytostatic	PS*	Torrance et al. 1975.
cnicin (41)	antimicrobial	Bordetella bronchiseptica Brucella abortus Pseudomonas aeruginosa	Vanhaelen-Fastre 1972, Vanhaelen-Fastre & Vanhaelen 1976.
	cytostatic	Staphylococcus aureus KB LE PS H	Vanhaelen-Fastre 1972, Vanhaelen-Fastre & Vanhaelen 1976, Gonzalez et al. 1978.
guaiene (25)	antimicrobial	Staphylococcus aureus a lipophilic diphtheroid	Morris et al. 1979.
longifolene (28)	antifungal	Boletus variegatus Rhizopogon roseolus	Melin & Krupa 1971.

Table 5 (continued)

Compound	Effect	Affected Organism	Reference
parthenin (44)	antifungal	*Sclerospora graminocola*	Char & Shankarabhat 1975.
parthenolide (45)	antifungal	*Cryptococcus albidus*	Blakeman & Atkinson 1979.
		Colletotrichum acutatum	
		Botrytis cinerea	
	antimicrobial	*Bacillus cereus*	Blakeman & Atkinson 1979.
		B. subtilis	
		B. megaterium	
		Micrococcus lysodeikticus	
		Mycobacterium spp.	
		and others	
	cytostatic	WI	Lee et al. 1971.
		H. EP-2	
		W-18 Va 2	
TRITERPENES			
oleanolic acid (59)	antifungal	*Saccharomyces carlsbergensis*	Anisimov et al. 1979.
MIXED TERPENES			
adhumulone (60)	antimicrobial	*Lactobacillis plantarum*	Hough et al. 1957.
		L. pastorianus	
cannabidiol (63)	antimicrobial	gram positive bacteria	Mechoulam & Edery 1973.
cannabigerol (64)	antimicrobial	gram positive bacteria	Mechoulam & Gaoni 1965, Mechoulam & Edery 1973.

Table 5 (continued)

Compound	Effect	Affected Organism	Reference
cohumulone (66)	antimicrobial	*Lactobacillus plantarum* *L. pastorianus*	Hough et al. 1957.
colupulone (67)	antimicrobial	same as above	Hough et al. 1957.
humulone (68)	antimicrobial	gram positive bacteria *Lactobacillus planatarum* *L. pastorianus*	Hansens 1947, Hough et al. 1957.
lupulone (66)	antimicrobial	gram positive bacteria	Hansens 1947.

*H.EP-2—Human epidermoid carcinoma of the larynx; KB—Human epidermoid carcinoma of the nasopharynx. Cell cultures; PS—P-388 lymphocytic leukemia; H—HeLa human carcinoma. Mouse; LE—L-1210 lymphoid leukemia. Intramuscular; WI—38—Human diploid fibroblasts; W-18Va2—Simian virus 40—transformed cell of human origin.

2.3 *Phytotoxins and Plant Growth Inhibitors*

The gibberellins and abscisic acid are terpenoid hormones that regulate physiological processes in plants but are not stored in or secreted from glandular hairs in high concentrations as the other terpenoids. The compartmentalization of the terpenes in the trichomes arising from the epidermis, isolated from other tissues, reduces the likelihood that they function as hormones or metabolic regulators; however, some of these compounds do affect the growth and development of plant cells and tissues (Table 6).

The volatile monoterpenes can vaporize from plants (Muller 1965a, Tyson et al. 1974) and interfere with normal physiological processes. Camphene (6), camphor (7), cineole (8), and α-pinene (12), all occurring in *Salvia leucophylla* Greene, *S. apiana* Jepson, and *S. mellifera* Greene, inhibited the growth of cucumber, *Cucumis sativus* L., with camphor (7) and cineole (8) being the most active (Muller and Muller 1964). The combined volatiles significantly reduced hypocotyl, radicle and lateral root growth and retarded respiration rates in the roots (Muller et al. 1968). Cineole (8) vapors retarded the respiration of *Avena fatua* L. roots and *Bromus rigidus* Mey. seedlings. Microscopic examination of the damaged cucumber tissues showed severely inhibited cell division and elongation of individual cells (Muller 1965b, Muller and Hauge 1967).

Other species whose growth is sensitive to cineole (8) are *Hordeum leporinum* Link, *Festuca megalura* Nutt. (del Moral and Muller 1970) and *Madia sativa* Molina (Halligan 1975); the latter is also affected by camphor (7) and the sesquiterpene caryophyllene (21).

Germination is sometimes influenced by the volatiles camphor (7), cineole (8) and caryophyllene (21); all retarded *Madia sativa* (Halligan 1975), whereas camphor (7), cineole (8) and α-pinene (12) inhibited radish, *Raphanus sativus* (Asplund 1968).

The sesquiterpene lactones, cumambrin-A (42) and -B (43) in the glands of *Artemisia nova* Nelson show various degrees of phytotoxicity when assayed with cucumber (McCahon et al. 1973) and several species (Table 5) from a tropical zone in Mexico (Amo and Anaya 1978, Anaya and Amo 1978). Parthenin (44), in addition to being antifungal and allergenic, is phytotoxic to beans, *Phaseolus vulgaris* L. (Garciduenas et al. 1972, Garciduenas and Dominguez 1976) and the crop plant *Eleusine coracana* (Linn.) Gaertn. (Kanchan 1975).

These phytotoxic compounds can escape from the plants into the surrounding environment by volatilization, leaching, exudation, or decomposition of the plant tissues (Rice 1974). If the compounds are present in sufficient concentration, they can affect the germination and growth of associated species or the species from which they originated. In this manner, chemicals originating from trichomes could influence ecological phenomena such as community composition, plant spacing and the rate of species replacement during succession (Rice 1974, Harborne 1977).

2.4 *Terpenoid Concentration in Plant Tissues and Their Active Concentration*

As noted in the previous discussions, the chemical compositions of the glandular fluids or their exudates are usually complex mixtures of chemicals in which the concentrations of individual compounds and classes of compounds vary considerably. Within a given species the quantitative and qualitative composition of the glands is dependent upon the plant's genetic makeup and geographic origin (Turner et al. 1977, 1978). Other factors influencing concentration are the organs on which the glands develop, the age of the organ and its physiological status (Table 7) (Guenther 1949, Henderson et al. 1970, Turner et al. 1977, 1978).

The terpenoid concentrations associated with trichomes can range from near 0 to 20% of the plant dry weight (Table 7). An important factor to emphasize, with regard to biological activities, is that these compounds are not diluted by distribution throughout the plant's tissues. They accumulate on the outside of the plant, either remaining within the glands or being secreted or exuded onto the epidermal surface. This external compartmentalization helps to provide high concentrations on the plant surface while eliminating the need for every cell to synthesize these compounds.

Having a mixture of natural products within the glands could bring added benefits such as synergism, where the activity of one compound is enhanced when in the presence of another compound or group of compounds (Harborne 1977). Also, the number of organisms affected can be increased with mixtures, since single compounds are often species specific in their activities. These possibilities make it difficult to determine the concentration necessary for biological activity.

The resin of *Hymenaea courbaril* discussed previously is composed of sesquiterpene hydrocarbons found in trichomes of other plants. When this resin was incorporated into the diet of the beet armyworm, *Spodoptera exigua*, at natural levels (0.50, 1.60, 3.20%) it increased larval susceptibility to disease, lengthened time to pupation and decreased pupal weight. It also provided a deterrent effect to treated tissues in palatability trials (Stubblebine and Langenheim 1977). Sesquiterpene lactones that deter insect feeding are active at concentrations of 1% (Burnett et al. 1974, Picman et al. 1978).

Like all antibiotics, the antimicrobial and antifungal effects of terpenoids are greatly dependent upon the assay organism, but many compounds are active at concentrations

Table 6. Terpenoid phytotoxins and plant growth inhibitors from glandular trichomes

Compound	Inhibition	Affected Species	Reference
MONOTERPENES			
camphene (6)	growth*	*Cucumis sativus* L.	Muller & Muller 1964.
camphor (7)	germination	*Madia sativa* Molina *Raphanus sativus* L.	Asplund 1968, Halligan 1975.
	growth	*Cucumis sativus* *Madia sativa*	Muller & Muller 1964, Halligan 1975.
cineole (8)	germination	*Madia sativa* *Raphanus sativus*	Asplund 1968, Halligan 1975.
	growth	*Bromus rigidus* Mey. *Cucumis sativus* *Festuca megalura* Nutt. *Hordeum leporinum* Link *Madia sativa*	Muller & Muller 1964, Muller et al. 1968, del Moral & Muller 1970, Halligan 1975.
	respiration	*Avena fatua* L. *Cucumis sativus*	Muller et al. 1968.
α-pinene (12)	germination	*Raphanus sativus*	Asplund 1968.
	growth	*Cucumis sativus*	Muller & Muller 1964.

Table 6 (continued)

Compound	Inhibition	Affected Species	Reference
SESQUITERPENES			
caryophyllene (21)	germination growth	Madia sativa Madia sativa	Halligan 1975. Halligan 1975.
cumambrin-A (42)	germination	Achyranthes aspera L. Bidens pilosa L. Crusea calocephala D.C.	Amo & Anaya 1978.
	growth	Ambrosia cumanensis H.B.K. Achyranthes aspera Bidens pilosa Crusea calocephala Cucumis sativus Mimosa pudica L.	Amo & Anaya 1978, McCahon et al. 1973.
cumambrin-B (43)	same as cumambrin-A above		
parthenin (44)	growth	Eleusine coracana (Linn.) Gaertn. Phaseolus vulgaris L.	Garciduenas et al. 1972, Kanchan 1975, Garciduenas & Dominguez 1976.
	respiration	P. vulgaris	Garciduenas & Dominguez 1976.

*The inhibitory effect may be observed in the growth of the roots, the hypocotyl, or the entire seedling.

Table 7. The concentration of glandular terpenoids in plant tissues

Plant	Terpenoid Type	Organ	Yield (% Dry Wt.)	Reference
Artemisia nova	monoterpenes	leaves and twigs	1.0–2.0	Nagy 1966.
	sesquiterpene lactones	inflorescence leaves	0.2–1.0 2.0–3.5	Kelsey & Shafizadeh 1980.
Beyeria viscosa	di- and triterpenes	leaves	20.0	Dell & McComb 1978b.
Cannabis sativa	cannabinoids	young bracts mature bracts young floral leaves mature floral leaves	4.0–7.3 3.3–6.2 3.7–5.7 2.0–5.6	Turner et al. 1977.
Eremophila fraseri	diterpenes and flavonoids	leaves	17.0	Dell 1975, Dell & McComb 1978b.
Mentha piperita	mono- and sesquiterpenes	leaves	0.3–2.0 (fresh wt.)	Guenther 1949.
Nepeta cataria	mono- and sesquiterpenes	leaves, stems and flowers	0.2 (fresh wt.)	Regnier et al. 1968.

Table 7 (continued)

Plant	Terpenoid Type	Organ	Yield (% Dry Wt.)	Reference
Newcastelia viscida	di- and triterpenes, also flavonoids	leaves	15.0	Dell & McComb 1975, 1978b.
Parthenium hysterophorus	sesquitepene lactones	capitulum leaves	8.0 5.0	Rodriguez et al. 1976a.
Pogostemon cablin	sesquiterpenes	primordial leaves fully developed leaves	12.0 1.0	Henderson et al. 1970.

between 10 (.001%) and 500 μg ml^{-1} (.05%) (Hough et al. 1957, Vanhaelen-Fastre 1972, Char and Shankarabhat 1975, Vanhaelen-Fastre and Vanhaelen 1976, Lee et al. 1977, Blakeman and Atkinson 1979, Morris et al. 1979). Parthenolide (45) at 12 μg ml^{-1} or less, inhibited the growth of *Bacillus cereus* and *B. subtilis* for 144 hours, *Micrococcus lysodeikticus* for 48 hours and both *Mycobacterium* spp. and *Nocardia* spp. for 24 hours. This concentration also inhibited the germination and germ tube growth of the filamentous fungus *Botrytis cinerea* (Blakeman and Atkinson 1979). Many sesquiterpene lactones are cytostatic at concentrations less than 1.0 μg ml^{-1} (.0001%) (Lee et al. 1971).

The volatile monoterpenes, because of their volatility, are more difficult to quantify and are often tested in a vapor phase (Eisner 1964, Muller and Muller 1964, Cobb et al. 1968, Muller et al. 1968, Melin and Krupa 1971, DeGroot 1972, Halligan 1975, Vasechko 1978). Asplund (1968) reported that camphor (7), α-pinene (12) and cineole (8) inhibited radish seed germination by 50% at concentrations of 0.5, 4.1, and 12.0 μg ml^{-1} respectively. Parthenin (44) at 50 and 100 μg ml^{-1} retarded root and hypocotyl growth of beans (Garciduenas et al. 1972, Garciduenas and Dominguez 1976), whereas cumambrin-A (42) and -B (43) were most inhibitory to the growth of native Mexican plants at 250 μg ml^{-1} (Amo and Anaya 1978).

3. PHENOLICS AND MISCELLANEOUS CONSTITUENTS

Information regarding phenolic compounds secreted by trichomes is not nearly as extensive as that for the terpenoids, even though phenolics appear to be quite common constituents of plant glands. Uphof stated in his 1962 review on plant hairs that "tannin is in mature glands almost always present", basing this statement on the traditional histochemical method for tannin staining by formation of colored complexes with iron salts. Formation of colored compounds with iron is a general property common to many phenolic substances not limited to tannins and has occasionally led to an overly broad definition of tannins (Ribereau-Gayon 1972). The work cited by Uphof would better be interpreted as indicative of the widespread occurrence of phenolic compounds in plant glands. A summary of the phenolic constituents isolated from trichomes (excluding flavonoids) which have reported biological activity is given in Table 8.

There are some species in which polymeric phenolics with tannin-like activity have been described as principal components of the glandular hair exudate. The surfaces of young leaves of the creosote bush *Larrea tridentata* (Zygophyllaceae) are covered with a sticky resin which hardens as the leaves age. The resin originates from single-cell trichomes which first develop a

Table 8. Biologically active phenolic substances stored or secreted from trichomes

Substances	Source	Biological Activity	References
Tannins	many species	astringent anti-feedant, reduces digestibility of proteins, allelopathy.	Uphof 1962; Feeny 1968, 1969, 1976; Rice 1974.
Phenol oxidase enzymes and phenol substrates	*Solanum poladenium* (Solanaceae) *Larrea* sp. (Zygophyllaceae)	polymerizes on mouthparts and tarsi restricting feeding; forms polymers with tannin-like activity.	Gibson 1971, 1976a, 1976b; Tingey and Gibson 1978; Thompson et al. 1979; Mabry al. 1977; Rhoades 1977
Nordihydroguaiaretic acid (lignan catechol)	*Larrea* sp. (Zygophyllaceae)	antioxidant; tannin-like activity.	Thompson et al. 1979; Mabry et al. 1977; Rhoades 1977
Phenolic stilbenes	*Alnus* sp. (Betulaceae)	fungicide, insecticide	Asakawa et al. 1977; Asakawa 1970, 1971, 1972; Suga et al. 1972; Overeem 1976
Phloroglucinol derivatives	*Dryopteris* sp. (Polypodiaceae)	anthelmintic	Huure et al. 1979; Widen & Britton 1971
Primin (benzoquinone derivative); other quinones	*Primula* sp. (Primulaceae)	contact allergen	Schildknect et al. 1967; Hausen 1978, 1980
Prenylated hydroquinone deratives	*Phacelia* sp. (Hydrophyllaceae)	contact allergens, radio-protective, cancer-protective	Reynolds and Rodriguez 1979, 1981; Reynolds et al. 1980.

suberin layer internal to the trichome primary wall, then senesce and collapses, resulting in extrusion of the cellular material through pores on the trichome onto the surface of the leaves (Thompson et al. 1979). Methylated flavonoids and lignans account for 80% of the resin constituents, of which the most abundant is a lignan catechol, nordihydroguaiaretic acid (NDGA) (73) which has been exploited commercially as a food anti-oxidant (Mabry et al. 1977). The phenolic aglycones including principally NDGA exhibit the tannin-like property of complexing with proteins and can thereby reduce the digestibility of the leaves. Rhoades (1977) has found a phenol oxidase system within the leaves which may enhance the tanning properties of the resin when leaves are macerated. Grasshoppers and moth larvae which feed on *Larrea* avoid the fresh resin by preferentially feeding on old leaves on which the resin is dry and hardened.

Glandular trichomes of the wild potato species *Solanum polyadenium S. berthaultii* and *S. tarijense* have recently become the subject of considerable study because of their effectiveness in protecting these species from insect damage, and because of the possibility of breeding this character into the cultivated potato *S. tuberosum* (Gibson 1971, 1976a, 1976b; Tingey and Gibson 1978). The glands contain a water-soluble mixture of phenolic compounds enclosed beneath the distended cuticle. The exudate is discharged when an insect ruptures the cuticle, and a polyphenol oxidase in the exudate causes polymerization of the exudate into a black viscous substance which hardens on mouthparts and tarsi and inhibits movement and feeding. The specific identity of the phenolic monomers in this system has not yet been reported.

Other glandular-pubescent plant species appear to have a similar method of insect resistance due to the sticky-viscid nature of their glandular exudate which impedes movement of small insects and arachnids. There are many species with mucilage-secreting trichomes that seem to function in this way (Schnepf 1974). Other species with similarly sticky secretions have received considerable study into the nature of their pest resistance but the chemistry of their glandular secretions have not been reported. Glandular haired species of tomato, *Lycopersicon* are resistant to infestation by spider mites and the greenhouse whitefly (Stoner 1970, Stoner et al. 1968, Georgiev and Satirova 1978), which are frequently seen to become trapped in the sticky exudate. The secretion of one highly resistant tomato strain is toxic when applied to the venter of the spider mite *Tetranychus* (Aina et al. 1972). Chemical analyses of tomato gland exudates have not been reported other than recent work by Isman and Duffey (1981) who found flavonoid glycosides to be a major

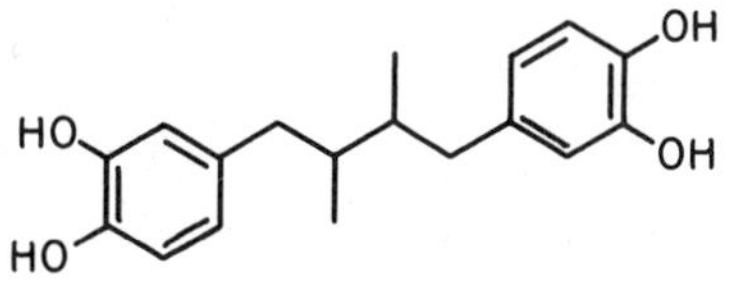

(73) nordihydroguariaretic acid from Larrea

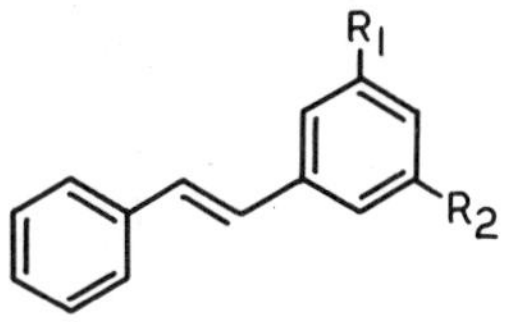

stilbenes from Alnus

(75) R_1 = H; R_2 = H
(76) R_1 = OH; R_2 = OH
(77) R_1 = OCH_3; R_2 = OH
(78) R_1 = OCH_3; R_2 = OCH_3

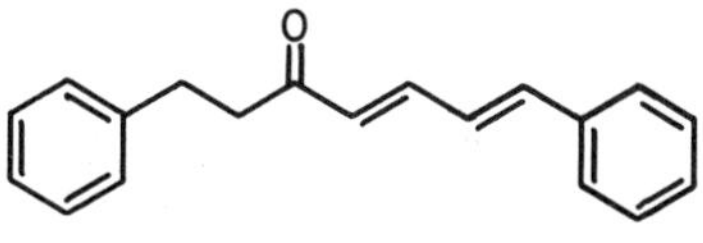

(74) alnustone from Alnus

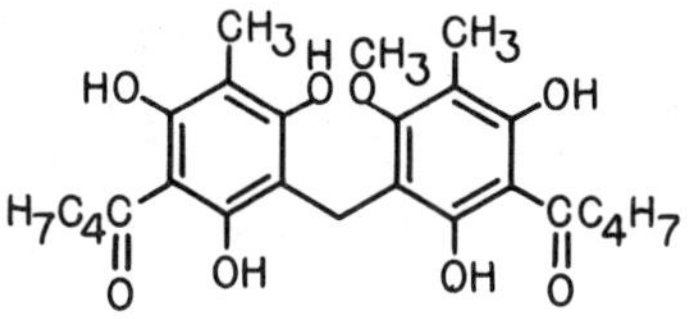

(79) margaspidin from Dryopteris

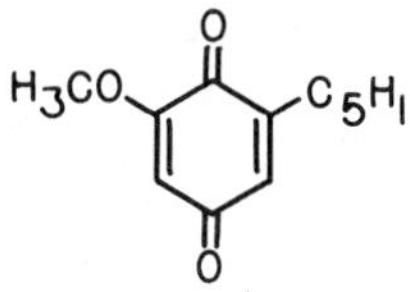

(80) primin from Primula

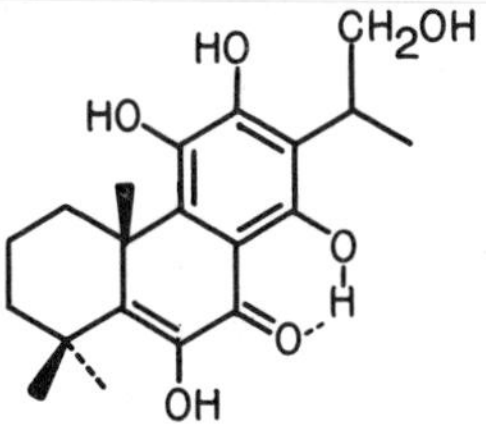

(81) coleone C from Coleus

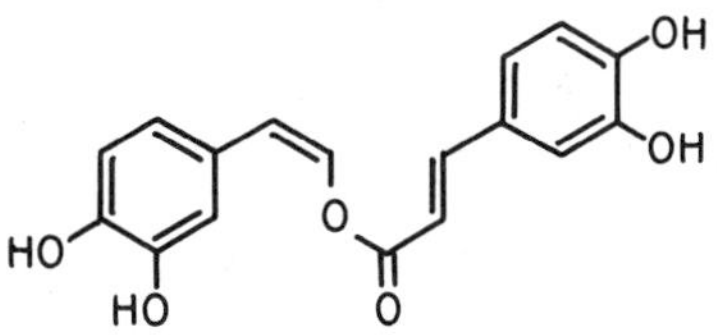

(82) from Plectranthus

Figure 7. Phenolic and miscellaneous constituents of glandular trichomes.

constituent of the trichomes and detrimental to larval growth of the fruitworm *Heliothis zea*.

Larvae of the alfalfa weevil *Hypera postica* and of the potato leaf hopper *Empoasca fabae*, both pests of the commercial alfalfa *Medicago sativa*, become trapped in the sticky exudate of glandular-haired *Medicago* species, and feeding and oviposition of adult weevils are inhibited. The resistance appears to be mechanical in nature, although the exudate of one species, *M. disciformis*, was toxic to alfalfa weevil larvae when applied topically (Shade et al. 1975, 1979; Johnson et al. 1980). Kreitner and Sorenson (1979a, 1979b) described two classes of glandular hairs on *Medicago*, procumbent and erect, which differ in morphology, distribution on the plant, presence on various species and varieties, and type of exudate. Only erect glands appeared to be effective in pest resistance. The exudate is lipophilic in histochemical tests, but has not been characterized further.

The winter buds of trees in the Betulaceae are covered by a viscous lipophilic material secreted by glandular trichomes (Asakawa et al. 1977). The exudate has been characterized in *Alnus sieboldiana* (Asakawa 1970, 1971, 1972) and *A. pendula* (Suga et al. 1972). It consists of a mixture of flavonoid aglycones, phenolic acids, esters and alcohols, neutral phenylpropanoid derivatives, rare diphenylheptanoid compounds such as alnustone (74), and neutral and phenolic stilbene derivatives (75-78) which are highly toxic to fungi and insects (Overeem 1976).

Ferns of the genus *Dryopteris* possess internal glandular trichomes which project into schizogenous spaces of the rhizomes and leaf base. A thick subcuticular layer of lipophilic excreta surrounds the single-celled bulbous head of each trichome. The excreta consists of a complex mixture of alkenes, even and odd-numbered fatty acids, and simple alkylated phenolics, the principal constituents being phloroglucinol derivatives which have anthelmintic activity (Huure 1979 et al., Huhtikangas et al. 1980, Widen and Britton 1971). There are a number of phloroglucinol derivatives which have been identified, one example being margaspidin (79) isolated from *D. marginalis*. These compounds are biosynthesized in the glandular cytoplast, excreted through the plasmalemma, and accumulated under the cuticle.

Quinonoid compounds are quite common throughout the plant kingdom, but only in few cases have they been reported in trichomes (Thomson 1971). This is not necessarily indicative of quinones being infrequent in trichomes; the extractions of quinones described in most of the reports in the phytochemical literature are from ground material of whole plant organs. The allergenic principle of *Primula obconica* (Primulaceae) which is responsible for most botanical contact dermatitis in Europe is a

benzoquinone, primin (80), first isolated by Nestler in 1904 and
identified by Schildknecht et al. in 1967. The quinone is stored
in very small non-capitate trichomes which release their contents
when broken, such as by persons handling the plant. Unlike
contact with stinging hairs there is no sensation on contact with
Primula. A first contact with the plant initiates development of
sensitization to primin, but a skin reaction at this time may not
appear. A second contact with the allergen occurring a week to
many months later may then cause a severe blistering rash which
only begins to develop hours to days after contact. Hausen
(1978) has screened the family Primulaceae for quinones and has
found primin and other unidentified quinones in numerous other
species. Whether or not the quinones of these other species were
stored in trichomes was not determined. In another report Hausen
(1980) cited two unidentified quinones as the cause of allergic
contact dermatitis from the orchids of the genus *Cypripedium*.
He suspects that these agents are stored in small trichomes on
the leaves, petals, and stems.

A series of diterpenoid quinones have been isolated from
glands on the surfaces of leaves and inflorescences of several
species of East-African Labiatae by members of Eugster's
laboratory in Zurich (Arihara et al. 1975a). A highly colored
quinone, Coleone C (81), was isolated from *Coleus aquaticus*,
a species with long spreading glandular hairs on the
inflorescence (Ruedi and Eugster 1971). The biological activities
of these compounds were not discussed. From similar glands
in *Plectranthus caninus* (Labiatae) dopaldehyde was isolated
in the form of its caffeic acid ester (82) (Arihara et al.
1975b). This was the first isolation from plants of dopaldehyde,
which is presumed to play a central role in the biogenesis
of opium alkaloids.

Recently a series of prenylated phenolic and quinonoid
compounds have been isolated from the viscid capitate-glandular
trichomes of *Phacelia*, a large genus in the Hydrophyllaceae of
herbs and subshrubs restricted to North America and the Andes.
The glandular exudate of several of these species causes many
persons severe allergic contact dermatitis (Munz 1934, Berry
et al. 1962). *Phacelia crenulata*, a springtime annual which
often grows profusely along desert roadsides in the southwestern
United States and northern Mexico, is responsible for most
clinical cases of *Phacelia* dermatitis. The principal trichome
constituents are geranylhydroquinone (83) and in lesser amount
geranylbenzoquinone (84) (Reynolds and Rodriguez 1979). Both
compounds are new natural products from plants, but previously
have been found in urochordates (Fenical 1974). Geranylhydroquinone has been synthesized as a drug and is found to have
radio-protective and cancer-preventive properties in experimental
animals (Rudali and Menetrier 1967, Lefevre et al. 1964).
Phacelia ixodes is a robust annual endemic to the Pacific

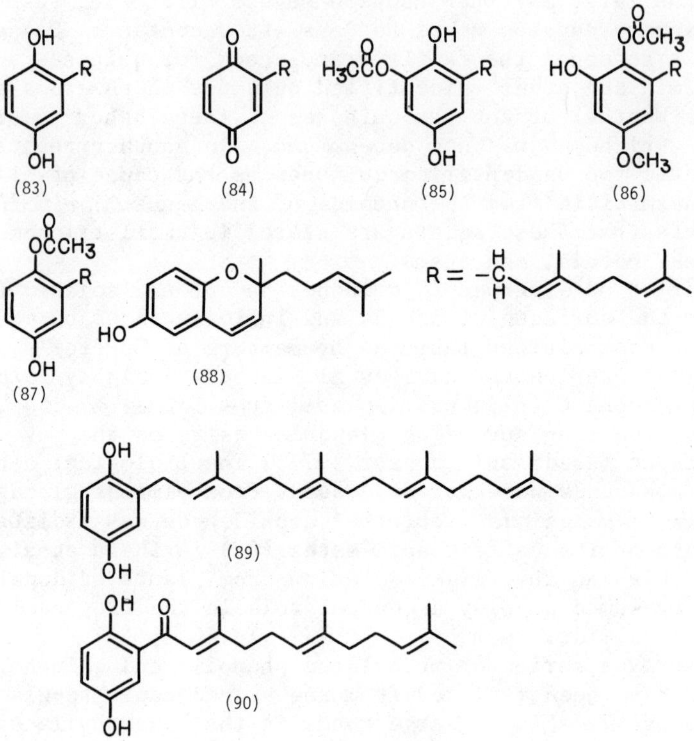

Figure 8. Phenolics from glandular trichomes of *Phacelia*.

seacoast of Baja California. The principal trichome constituents include geranylhydroquinone and geranylbenzoquinone, and also the acetate and methyl substituted derivatives 85-87 as well as the chromene derivative of geranylhydroquinone 88. Substances 85-87 are newly reported natural products (Reynolds and Rodriguez, 1981a), while the chromene 88 has been isolated from heartwood of the tropical tree *Cordia alliodora* in the family Boraginaceae (Manners and Jurd 1977). Giving the trichomes a granular appearance are precipitates of 6,7-dimethoxyapigenin in the glandular exudate. This flavone has previously been found as the aglycone only in the family Labiatae (Ulubelen et al. 1979, Brieskorn and Biechele 1969).

Phacelia minor and *P. parryi* are closely related species widespread through the coastal mountains of southern California, with *P. parryi* extending into Baja California. The trichome constituents of each are the same, the major one being geranylgeranylhydroquinone (89), and a lesser one being 2-(1-oxofarnesyl)-hydroquinone (90) (Reynolds and Rodriguez 1981b). Compound 89 is newly reported in plants, but has been reported earlier in marine sponges (Cimino et al. 1972). Constituent 90 is a new natural product; the parent structure, farnesylhydroquinone, has been reported both in a brown alga (Ochi et al. 1979) and *Wigandia kunthii* of the family Hydrophyllaceae (Gomez et al. 1980).

Guinea pigs sensitized to crude trichome extracts were used to assay these compounds for allergenic potential, and one human sensitization study was conducted with geranylhydroquinone (Reynolds et al. 1980). Metabolites 83, 84, 86, 89, and 90 proved to be potent contact allergens. The mechanism by which phenolic haptens form active antigens in allergic contact dermatitis has not yet been established. One study has suggested that an oxidation to the quinone form takes place *in vivo* followed by nucleophilic substitution on the ring by the sulfhydryl or amino group of a cell-surface protein in the skin. The haptenprotein complex thus formed is then recognized as foreign by the cellular immune system which then acts against those cells to which the antigen is bound (Byck and Dawson 1968). Cross-sensitivity reactions were observed between substances (83), (84), and (89), and between (89) and (90). The mechanism described above would account for the cross-sensitivity of (83) and (84). A similar explanation for the cross-sensitivity of (89) and (90) could be imagined if (90) were in the *enol* form resulting from the transfer of a proton from the phenolic hydroxyl to the carbonyl. A similar mechanism to explain the activity of substance (86) requires first a deacetylation *in vivo*, followed by oxidation to the ortho-quinone before a nucleophilic substitution by protein could occur. Lack of activity by compound (85) may be due to steric hindrance by the acetoxyl group preventing substitution at the adjacent position of the ring. A similar situation was

seen in sesquiterpene lactones (Epstein et al. 1980). The
placement of an acetoxyl group adjacent to the site susceptable to
nucleophilic addition nullified the activity of an otherwise
active contact allegen.

There are few reports of alkaloids being significant
constituents of glandular trichome secretions, except in stinging
hairs. Several *Nicotiana* species and *Petunia* species
(Solanaceae) are highly resistant to aphids, spider mites,
and larvae of tobacco hornworm due to alkaloids in the gummy
exudate of the glandular hairs (Thurston et al. 1966, Parr
and Thurston 1968, Thurston 1970, Patterson et al. 1974).
The major alkaloid is nicotine, while nornicotine and anabasine
are present in lesser amounts in some species. It was shown
that trichome delivery of alkaloids is critical to effective
resistance to tobacco hornworm (Thurston 1970). Tobacco hornworm
larvae can rapidly eliminate from their bodies high levels
of nicotine ingested, injected, or applied topically and
consequently are able to survive on strains of *N. tabacum*
with high nicotine content (Self et al. 1964). However, the
same larvae could not survive on species of *Nicotiana* that
possess nicotine containing glandular hairs. Apparently
continuous contact with nicotine in the trichome exudate exceeds
the ability of the larvae to tolerate the alkaloid. Washing
the trichome exudate from the leaf surface of resistant species
reduced or eliminated toxicity.

In a study of the localization of alkaloids in *Vinca
herbacea* (Apocynaceae) using histochemical methods and
fluorescence microscopy, Kartmazova and Lyapunova (1976)
detected unspecified alkaloids in the trichomes, as well as
in other secretory cells and latex vessels. In a study of
the alkaloids of *Mucuna pruriens* (Fabaceae), a species with
irritant trichomes on the seed pods, Ghosal et al. (1971)
reported six indole alkyl amines present in all parts of the
plant except in the pod trichomes which yielded only serotonin.

4. CONCLUSIONS

A review of the literature and our research findings
indicate that a large number of secondary metabolites produced
and/or stored in glandular plant trichomes exhibit potent
biological activities. These substances have been shown to
be associated with anti-tumor, cytotoxic, anti-microbial and
phytotoxic activity. They are also known to poison livestock,
to act as insect feeding deterrents and to cause allergic
contact dermatitis in humans. Although higher plants are capable
of producing literally thousands of secondary constituents,
it is interesting to note that the trichomes generally contain
the most active constituents. Also, glandular trichomes contain
less amounts and kinds of secondary constituents as compared

to the whole leaf or other plant tissues. Therefore, the idea that trichomes are indeed the "first line of defense" is supported by studies by entomologists and phytochemists. The preponderance of terpenoids in trichomes suggests that these compounds are very effective against pathogens and herbivores, but only a few plant families have been investigated, so it is likely that other classes of secondary metabolites will be detected that are effective against pathogens and phytophagous insects. Nevertheless, a considerable amount of research is still needed to elucidate the structures of constituents secreted and/or stored in hairs in order to appreciate their biological activities and functions. Also, physiological studies coupled with phytochemical investigations will be necessary if we wish to understand the adaptive significance of secondary metabolites in trichomes of plants exposed to harsh environmental pressures such as drought and extreme radiation.

ACKNOWLEDGEMENTS

R.K. wishes to thank the National Science Foundation (PFR 78-826314) for financial assistance. E.R. thanks the National Science Foundation (PCM 8209100) and the National Institute of Health (AI-18398-01A1) for financial support.

5. REFERENCES

Abrams, L., 1940, 1944, 1951, 1960, "Illustrated Flora of the Pacific States-Washington, Oregon, and California," Stanford University Press, Stanford.

Aina, O.J., Rodriguez, J.G., and Knavel, D.E., 1972, Characterizing resistance to *Tetranychus urticae* in tomato. J. Econ. Entomol. 65:641.

Amelunxen, F., and Arbeiter, H., 1969, Untersuchungen an den Drüsenhaaren von *Cleome spinosa* L. Z. Pflanzenphysiol. 61:73.

Amelunxen, F., Wahlig, T., and Arbeiter, H., 1969, Über den Nachweis des ätherischen Öls in isolierten Drüsenhaaren und Drüsenschuppen von *Mentha piperita* L. Z. Pflanzenphysiol. 61:68.

Amo, S.D., and Anaya, A.L., 1978, Effect of some sesquiterpenic lactones on the growth of certain secondary tropical species. J. Chem. Ecol. 4:305.

Anaya, A.L., and Amo, S.D., 1978, Allelopathic potential of *Ambrosia cumanensis* H.B.K. (Compositae) in a tropical zone of Mexico. J. Chem. Ecol. 4:289.

Anisimov, M.M., Shcheglov, V.V., Strigina, L.I., Chetyrina, N.S., Uvarova, N.I., Oshitok, G.I., Alad'ina, N.G., Vecherko, L.P., Zorina, A.D., Matyukhina, L.G., and Saltykova, I.A., 1979, Chemical structure and antifungal activity of a number of triterpenoids. Izv. Akad. Nauk SSSR, Ser. Biol., 570.

Arihara, S., Ruedi, P., and Eugster, C.H., 1975a, Neue spiro-cyclopropyl-cyclohexendion-diterpene: Coleone M, N, P, Q, R sowie Barbatusin aus *Plectranthus caninus* Roth und Coleone O aus *Coleus somaliensis* S. Moore. Helv. Chim. Acta 58:343.

Arihara, S., Ruedi, P., and Eugster, C.H., 1975b, Dopaldehyd: erstmalige Isolierung aus einer Pflanze in Form seines (Z)-Enol-(E)-Kaffeesäureesters. Helv. Chim. Acta 58: 447.

Asakawa, Y., 1970, Chemical Constituents of *Alnus firma* (Betulaceae). I. Phenyl propane derivatives isolated from from *Alnus firma*. Bull. Chem. Soc. Japan 43:2223.

Asakawa, Y., 1971, Chemical constituents of *Alnus sieboldiana* (Betulaceae). II. The isolation and structure of flavonoids and stilbenes. Bull. Chem. Soc. Japan 44:2761.

Asakawa, Y., 1972, Chemical constituents of *Alnus sieboldiana* (Betulaceae). III. The synthesis and stereochemistry of yashabushiketols. Bull. Chem. Soc. Japan 45:1794.

Asakawa, Y., Takemoto, T., Wollenweber, E., and Aratani, T., 1977, Lasiocarpin A, B and C, three novel phenolic triglycerides from *Populus lasiocarpa*, Phytochem. 16:1791.

Asplund, R.O., 1968, Monoterpenes: Relationship between structure and inhibition of germination. Phytochem. 7:1995.

Bakker, H.J., Ghisalberti, E.L., and Jefferies, P.R., 1972, Biosynthesis of diterpenes in *Beyeria leschenaultii*. Phytochem. 11:2221.

Berry, C.Z., Shapiro, S.I., and Dahlen, R.F., 1962, Dermatitis venenata from *Phacelia crenulata*, Arch. Dermatol. 85:737.

Birch, A.J., Brown, W.V., Corrie, J.E.T., and Moore, B.P., 1972, Neocembrene-A, a termite trail pheromone. J. Chem. Soc. Perkin I, 2653.

Birch, A.J., Subba Rao, G.S.R., and Turnbull, J.P., 1966, Eremolactone, Tetrahedron Lett. No. 39:4749.

Blakeman, J.P., and Atkinson, P., 1979, Antimicrobial properties and possible role in host-pathogen interactions of parthenolide, a sesquiterpene lactone isolated from glands of *Chrysanthemum parthenium*. Physiol. Plant Path. 15:183.

Brieskorn, C.H., and Biechele, W., 1969, 6-Methoxygenwanin-Ein weiters flavon aus Labiaten, Tetrahedron Lett. 13:2603.

Buchanan, M.A., 1963, Extraneous components of wood. *In:* Browning, B.L. (ed.), The Chemistry of Wood, 313, John Wiley & Sons, New York, London, Sydney.

Burnett, W.C., Jr., Jones, S.B., Jr., Mabry, T.J., and Padolina, W.G., 1974, Sesquiterpene lactones--insect feeding deterrents in *Vernonia*. Biochem. Syst. Ecol. 2:25.

Buttery, R.G., and Ling, L.C., 1967, Identification of hop varieties by gas chromatographic analysis of their essential oils. J. Agr. Food Chem. 15:531.

Buttery, R.G., Lundin, R.E., and Ling, L., 1967, Characterization of some C_{15} constituents of hop oil. J. Agr. Food Chem. 15:58.

Byck, J.S., and Dawson, C.R., 1968, Assay of protein-quinone coupling involving compounds structurally related to the active principle of poison ivy. Anal. Biochem. 25:123.

Cade, A.R., 1954, Essential oils. *In:* Reddish, G.F. (ed.), Antiseptics, Disinfectants, Fungicides, and Chemical and Physical Sterilization, 279, Lea & Febiger, Philadelphia.

Char, N.B.S., and Shankarabhat, S., 1975, Parthenin: A growth inhibitor behavior in different organisms. Experientia 31:1164.

Ciminio, G., De Stefano, S., and Minale, L., 1972, Prenylated quinones in marine sponges: *Ircinia* sp., Experientia 28:1401.

Clayton, R.B., 1970, The chemistry of hormonal interactions. Terpenoid compounds in ecology. *In:* Sondheimer, E. and Simeone, J.B. (eds.), Chemical Ecology, 235, Academic Press New York-London.

Coates, P., Ghisalberti, E.L., and Jefferies, P.R., 1977, The chemistry of *Eremophila* spp. VIII. A cembrenetriol from *E. clarkei*. Aust. J. Chem. 30:2717.

Cobb, F.W., Jr., Krstic, M., Zavarin, E., and Barber, H.W., Jr., 1968, Inhibitory effects of volatile oleoresin components on *Fomes annosus* and four *Ceratocystis* species. Phytopath. 58:1327.

Croft, K.D., Ghisalberti, E.L., Jefferies, P.R., Raston, C.L., White, A.H., and Hall, S.R., 1977, The chemistry of *Eremophila* spp. - VI., Stereochemistry and crystal structure of dihydroxyserrulatic acid. Tetrahedron 33:1475.

Croteau, R., 1977, Site of monoterpene biosynthesis in *Majorana hortensis* leaves. Plant Physiol. 59:519.

Dawson, R.M., Jarvis, M.W., Jefferies, P.R., Payne, T.G., and Rosich, R.S., 1966, Acidic constituents of *Dodonaea lobulata*. Aust. J. Chem. 19:2133.

DeGroot, R.C., 1972, Growth of wood-inhibiting fungi in saturated atmospheres of monoterpenoids. Mycologia 64:863.

del Moral, R., and Muller, C.H., 1970, The allelopathic effects of *Eucalyptus camaldulensis*. Am. Midl. Nat. 83:254.

Dell, B., 1975, Geographical differences in leaf resin components of *Eremophila fraseri* F. Muell. (Myoporaceae). Aust. J. Bot. 23:889.

Dell, B., 1977, Distribution and function of resins and glandular hairs in western Australian plants. J. Roy. Soc. West. Aust., 59:119.

Dell, B., and McComb, A.J., 1975, Glandular hairs, resin production, and habitat of *Newcastelia viscida* E. Pritzel *(Dicrastylidaceae)*. Aust. J. Bot. 23:373.

Dell, B., and McComb, A.J., 1978a, Biosynthesis of resin terpenes in leaves and glandular hairs of *Newcastelia viscida*. J. Exp. Bot. 29:89.

Dell, B., and McComb, A.J., 1978b, Plant resins: Their formation, secretion and possible functions. Adv. Bot. Res. 6:277.

Doskotch, R.W., Fairchild, E.H., Huang, C.T., Wilton, J.H., Beno, M.A., and Christoph, G.G., 1980, Tulirinol, an antifeedant sesquiterpene lactone for the gypsy moth larvae from *Liriodendron tulipifera*. J. Org. Chem. 45: 1441.

Eisner, T., 1964, Catnip: Its raison d'etre. Science 146:1318.

Eisner, T., Johnessee, J.S., Carrel, J., Hendry, L.B., and Meinwald, J., 1974, Defensive use by an insect of a plant resin. Science 184:996.

Eisener, T., Kluge, A.F., Ikeda, M.I., Meinwald, Y.C., and Meinwald, J., 1971, Sesquiterpenes in the osmeterial secretion of a papilionid butterfly, *Battus polydamas*, J. Insect Physiol. 17:245.

Epstein, W.L., Reynolds, G.W., and Rodriguez, E., 1980, Cross-sensitivity in costus-sensitized patients. Arch. Dermatol. 1116:59.

Esau, K., 1965, Plant Anatomy, John Wiley and Sons, New York, London, Sydney.

Fahn, A., 1979, Secretory Tissues in Plants. Academic Press, London, New York.

Feeny, P.P., 1968, Effect of oak leaf tannins on larval growth of the winter moth *Operophtera brumata*, J. Insect Physiol. 14:805.

Feeny, P.P., 1969, Inhibitory effect of oak leaf tannins on the hydrolysis of proteins by trypsin. Phytochem. 8:2119.

Feeny, P.P., 1976, Plant apparency and chemical defense, *In:* J. Wallace and R. Mansell (eds.), "Recent Advances in Phytochemistry Vol. 10 - Biochemical Interaction between Plants and Insects" Plenum Press, New York.

Fenical, W., 1974, Geranylhydroquinone, a cancer-protective agent from the tunicate *Aplidium* sp., *In:* "4th Proceedings of Food-Drugs from the Sea", Marine Technological Society, Washington, D.C.

Garciduenas, M.R., Dominguez, X.A., Fernandez, J., and Alanis, G., 1972, New Growth inhibitors from *Parthenium hysterophorus*. Rev. Latinoamer. Quim. 2:52.

Garciduenas, M.R., and Dominguez, X.A., 1976, Partenina, achilina y eugarzasadina tres nuevos inhibidores lactonicos del desarrollo vegetal. Turrialba 26:10.

Georgiev, K., and Sotirova, V., 1978, A study of resistance of wild tomato species to the greenhouse whitefly *(Trialeurodes vaporarium* Westw.). Genet. Sel. 11:214.

Ghisalberti, E.L., Jefferies, P.R., and Sheppard, P., 1975, A new class of diterpenes from *Eremophila decipiens*. Tetrahedron Lett. No. 22 + 23:1775.

Ghisalberti, E.L., Jefferies, P.R., Knox, J.R., and Sheppard, P.N., 1977, The chemistry of *Eremophila* sp - VII. An epoxycembradienol from *Eremophila georgei*. Tetrahedron 33:3301.

Ghosal, S., Singh, S., and Bhattacharya, S.K., 1971, Alkaloids of *Mucuna pruriens:* Chemistry and pharmacology. Planta Med. 19:279.

Gibson, R.W., 1971, Glandular hairs providing resistance to aphids in certain wild potato species, Ann. Appl. Biol. 68:113.

Gibson, R.W., 1976a, Glandular hairs are a possible means of limiting aphid damage to the potato crop, Ann. Appl. Biol. 82:143.

Gibson, R.W., 1976b, Glandular hairs on *Solanum polyadenium* lessen damage by the Colorado beetle, Ann. Appl. Biol. 82:147.

Gomez, F., Quijano, L., Calderon, J.S., and Rios, T., 1980, Terpenoids isolated from *Wigandia kunthii,* Phytochem. 19:2202.

Gonzalez, A.G., Darias, V., Alonso, G., Boada, J.N., and Feria, M., 1978, Cytostatic activity of sesquiterpene lactones from *Compositae* of the Canary Islands. Planta Medica 33:356.

Goodwin, T.W., 1967, The biological significance of terpenes in plants. *In:* Pridham, J.B. (ed.), Terpenoids in Plants, 1-23. Academic Press, London-New York.

Guenther, E., 1949, The Essential Oils, 3. D. Van Nostrand Co., Inc., New York, Toronto, London.

Guenther, E., 1952, The Essential Oils, 6. D. Van Nostrand Co., Inc., New York, Toronto, London.

Halligan, J.P., 1975, Toxic terpenes from *Artemisia californica*. Ecology 56:999.

Hansens, M., 1947, Antiseptic activity of humulone and lupulone. Congr. Intern. Inds. Fermentation, Confs et Communs, 302. Chem. Abst. 1948, 42:7928a.

Harborne, J.B., 1977, Introduction to Ecological Biochemistry. Academic Press, London, New York, San Francisco.

Hausen, B.M., 1978, On the occurrence of the contact allergen primin and other quinoid compounds in species of the family Primulaceae. Arch. Derm. Res. 261:311.

Hausen, B.M., 1980, Allergic contact dermatitis to quinones in *Paphiopedilum hynaldianum (Orchidaceae),* Arch. Derm. 116:327.

Hausen, B.M., and Schulz, K.H., 1976, Chrysanthemum allergy III. Identification of the allergens. Arch. Derm. Res. 255: 111.

Hayashi, N., and Komae, H., 1977, The trail and alarm pheromones of the ant, *Pristomyrmex pungens* Mayr. Experientia 33:424.

Henderson, W., Hart, J.W., How, P., and Judge, J., 1970, Chemical and morphological studies on sites of sesquiterpene accumulation in *Pogostemon cablin* (patchouli). Phytochem. 9:1219.

Hendriks, H., Malingre, T.M., Batterman, S., and Bos, R., 1975, Mono- and sesqui-terpene hydrocarbons of the essential oil of *Cannabis sativa*. Phytochem. 14:814.

Hood, L.V.S., Dames, M.E., and Barry, G.T., 1973, Headspace volatiles of marijuana. Nature 242:402.

Hough, J.S., Howard, G.A., and Slater, C.A., 1957, Bacteriostatic activities of hop resin materials. J. Inst. Brew. 63:331.

Howse, P.E., 1975, Chemical defense of ants, termites and other insects: Some outstanding questions. Proc. IUSSI Symp. (Dijon), 23.

Howse, P.E., Baker, R., and Evans, D.A., 1977, Multifunctional secretions in ants. Proc. VIII. Int. Congr. IUSSI (Wageningen), 44.

Hrdy, I., Krecek, J., and Vrkoc, J., 1977, Biological activity of soldiers secretions in the termites: *Nasutitermes rippertii, N. costalis* and *Prorhinotermes simples*. Proc. VIII. Int. Congr. IUSSI (Wageningen), 303. Chem. Abst. 1978, 89: 194289.

Huhtikangas, A., Huurret, A., and Partanen, A., 1980, Electron microscopic investigations on internal glandular hairs of *Dryopteris assimilis* ferns: 2. Nature and origins of non-phloroglucinol constituents of the glandular excrete, Planta Med. 38:62.

Huure, A., Huhtikangas, A., and Widen, C.J., 1979, Electron microscopic investigations on internal glandular hairs of *Dryopteris* ferns: I. General observations on hair structure and excrete accumulation. Planta Med. 35:262.

Ishikawa, S., and Hirao, T., 1966, Studies on the olfactory sensations in the larvae of the silkworm, *Bombyx mori* (III). Attractants and repellents of hatched larvae. Bull. Sericult. Expt. Sta. Tokyo. 20:291.

Isman, M.B., and Duffey, S.S., 1981, Experientia (in press).

Jefferies, P.R., and Ratajczak, T., 1973, Isopimara-9(11),15-diene-3β, 19-diol from *Newcastlia viscida* (Verbenaceae). Aust. J. Chem. 26:173.

Johnson, K.J.R., Sorenson, E.L., and Horber, E.K., 1980, Resistance of glandular-haired *Medicago* species to oviposition by alfalfa weevils *(Hypera postica)*, Environ. Entomol. 9:241.

Kanchan, S.D., 1975, Growth inhibitors from *Parthenium hysterophorus* Linn. Cur. Sci. 44:358.

Karlson, P., 1970, Terpenoids in insects. *In:* Goodwin, T.W. (ed.), Natural Substances Formed Biologically from Mevalonic Acid, 145-156. Academic Press, London-New York.

Kartmazova, L.S., and Lyapunova, P.W., 1976, The anatomic structure and localization of alkaloids in the vegetative organs of the herbaceous periwinkle *(Vinca herbacea* Waldst. et Kit.). Farmatsiya (Mosc.) 25: 22; Bio. Abstr. 63:46422.

Kelsey, R.G., and Shafizadeh, F., 1979, Sesquiterpene lactones and systematics of the genus *Artemisia.* Phytochem. 18:1591.

Kelsey, R.G., and Shafizadeh, F., 1980, Glandular trichomes and sesquiterpene lactones of *Artemisia nova (Asteraceae).* Biochem. Syst. Ecol. 8:371.

Kreitner, G.L., and Sorenson, E.L., 1979a, Glandular trichomes on *Medicago* species. Crop Sci. 19:380.

Kreitner, G.L., and Sorenson, E.L., 1979b, Glandular secretory system of alfalfa species. Crop Sci. 19:499.

Lee, K.H., Huang, E.S., Piantadosi, C., Pagano, J.S., and Geissman, T.A., 1971, Cytotoxicity of sesquiterpene lactones. Cancer Res. 31:1649.

Lee., K.H., Ibuka, T., Wu, R.Y., and Geissman, T.A., 1977, Structure-antimicrobial activity relationships among the sesquiterpene lactones and related compounds. Phytochem. 16:1177.

Lefevre, R., Baranger, P., Fesneau, R., and Husson, M.J., 1964, Chimiotherapie anticancereuse, Acta: Unio. Internationalis Contra Cancrum 20:329.

Levin, D.A., 1973, The role of trichomes in plant defense, Quar. Rev. Biol. 48:1.

Lonkar, A., Mitchell, J.C., and Calnan, C.D., 1974, Contact dermatitis from *Parthenium hysterophorus.* Trans. St. John's Hosp. Derm. Soc. (London) 60:43.

Loomis, W.D., and Croteau, R., 1973, Biochemistry and physiology of lower terpenoids. *In:* Runeckles, V.C., and Mabry, T.J. (eds.), Terpenoids: Structure, Biogenesis, and Distribution, Recent Advances in Phytochemistry 6:147-185. Academic Press. New York-London.

Lüttge, U., 1971, Structure and function of plant glands, Ann. Rev. Plant Physiol. 22:23.

Mabry, T.J., Difeo, D.R., Jr., Sakakibara, M., Bohnstedt, C.G., Jr., and Siegler, D., 1977, The natural products of *Larrea,* Ch. 5 *In:* Mabry, T.J., Hunziker, J.H., and Difeo, Jr., D.R. (eds.), "Creosote bush, Biology and Chemistry of *Larrea* in New World Deserts", Dowden, Hutchinson, and Ross, Stroudsburg.

Malingre, M.Th., Smith, D., and Batterman, S., 1969, De isolering en gaschromatografische analyse van de vluchtige olie uit afzonderlijke klierharen van het labiatentype. Pharm. Weekbl. 104:429.

Malingre, Th., Hendriks, H., Batterman, S. Bos, R., and Visser, J., 1975, The essential oil of *Cannabis sativa*. Planta Medica 28:56.

Manners, G.D., and Jurd, L., 1977, The hydroquinone terpenoids of *Cordia alliodora*, J. Chem. Soc., Lond. Perkin I, 405.

Maruzzella, J.C., and Henry, P.A., 1958, The *in vitro* antibacterial activity of essential oils and oil combinations. J. Am. Pharm. Assoc., Sci. Ed. 47:294.

Maruzzella, J.C., and Liguori, L., 1958, The *in vitro* antifungal activity of essential oils. J. Am. Pharm. Assoc., Sci. Ed. 47:250.

Maruzzella, J.C., and Sicurella, N.A., 1960, Antibacterial activity of essential oil vapors. J. Am. Pharm. Assoc., Sci. Ed. 49:692.

Maslen, E.N., Raston, C.L., and White, A.H., 1977, Crystal Structure of an epoxycembradienol, 3,15-epoxy-4-hydroxycembra-7(Z), 11 (Z)-diene. Tetrahedron 33:3305.

McCahon, C.B., Kelsey, R.G., Sheridan, R.P., and Shafizadeh, F., 1973, Bull. Torrey Bot. Club 100:23.

Mechoulam, R., and Gaoni, Y., 1965, Hashish-IV. The isolation and structure of cannabinolic, cannabidiolic and cannabigerolic acids. Tetrahedron 21:1223.

Mechoulam, R., and Edery, H., 1973, Structure-activity relationships in the cannabinoid series. *In:* Mechoulam, R. (ed.), Marijuana. Chemistry, Pharmacology, Metabolism and Clinical Effects, 101. Academic Press, New York-London.

Melin, E., and Krupa, S., 1971, Studies on ectomycorrhizae of pine II. Growth inhibition of mycorrhizal fungi by volatile organic constituents of *Pinus silvestris* (scots pine) roots. Physiol. Plant. 25:337.

Michie, M.J., and Reid, W.W., 1968, Biosynthesis of complex terpenes in the leaf cuticle and trichomes of *Nicotiana tabacum*. Nature 218:578.

Morris, J.A., Khettry A., and Seitz, E.W., 1979, Antimicrobial activity of aroma chemicals and essentials oils. J. Am. Oil Chem. Soc. 56:595.

Morton, J., 1972, Further associations of plant tannins and human cancer, Quar. J. Crude Drug Res. XII:1829.

Muller, C.H., 1965a, Inhibitory terpenes volatilized from *Salvia* shrubs. Bull. Torrey Bot. Club 92:38.

Muller, W.H., and Muller, C.H., 1964, Volatile growth inhibitors produced by *Salvia* species. Bull. Torrey Bot. Club 91:327.

Muller, W.H., 1965b, Volatile materials produced by *Salvia leucophylla:* Effects on seedling growth and soil bacteria. Botan. Gaz. 126:195.

Muller, W.H., and Hauge, R., 1967, Volatile growth inhibitors produced by *Salvia leucophylla:* Effect on seedling anatomy. Bull. Torrey Bot. Club 94:182.

Muller, W.H., Lorber, P., and Haley, B., 1968, Volatile growth inhibitors produced by *Salvia leucophylla:* Effect on seedling growth and respiration. Bull. Torrey Bot. Club 95:415-422.

Munz, P.A., 1934, Dermatitis produced by *Phacelia (Hydrophyllacead),* Science 76:194.

Munz, P.A., 1959 "A California Flora", University of California Press, Berkeley, Los Angeles.

Mutton, D.B., 1962, Wood resins, *In:* Hillis, W.E. (ed.), Wood Extractives, 331-363. Academic Press, New York-London.

Nagy, J., 1966, Volatile oils and antibiosis of *Artemisia.* Ph.d. dissertation, Colorado State Univ.

Nestler, A., 1904, Hautreizende Primeln, Berlin: Gebr Borntrager.

Nicholas, H.J., 1973, Terpenes, *In:* Miller, L.P. (ed), Phytochemistry, 254-309. Van Nostrand Reinhold Co., New York.

Nigam, M.C., Handa, K.L., Nigam, I.C. and Levi, L., 1965, Essential oils and their constituents XXIX. The essential oil of marihuana: Composition of genuine Indian *Cannabis sativa* L. Can. J. Chem. 43:3372.

Noble, T.A., and Epstein, W.W., 1977, The absolute stereochemistry and corrected structure of the monoterpene ether from *Artemisia tridentata.* Tetrahedron Lett. 45:3931.

Ochi, M., Kotsuki, H., Inoue, S., Taniguchi, M., and Tokoroyama, T., 1979, Isolation of 2-(3,7,11-trimethyl-2,6,10-dodecatrienyl)-hydroquinone from the brown seaweed *Dictyopteris undulata.* Chem. Lett. 831.

Oshkaev, A. Kh., 1977, Toxic effect of monterpenes on conifer-needle-chewing [insects] and cone and seed pests. Izv. Vyssh. Uchebn. Zaved. Lesn. Zh. 20:28. Chem. Abst. 1978, 88:116-302.

Overeem, J.C., 1976, Pre-existing antimicrobial substances in plants and their role in disease resistance, pages 195-206 *In:* Friend, J., and Threlfall, D.R. (eds.), "Biochemical Aspects of Plant-Parasite Relationships", Academic Press. New York, London.

Parr, J.C. and Thurston, R., 1968, Toxicity of *Nicotiana* and *Petunia* species to larvae of the tobacco hornworm. J. Econ. Entomol. 61:1525.

Patterson, C.G., Thurston, R., and Rodriguez, J.G., 1974, Twospotted spider mite resistance in *Nicotiana* species. J. Econ. Entomol. 67:341.

Picman, A.K., Elliott, R.H., and Towers, G.H.N., 1978, Insect feeding deterrent property of alantolactone. Biochem. Syst. Ecol. 6:333.

Politis, M.J., 1946, Sur la formation des glucosides amers dans les poils glanduleux de certaines plantes. Compt. Rend. Acad. Sci. (Paris) 222:910.

Prestwich, G.D., 1979, Chemical defense by termite soldiers. J. Chem. Ecol. 5:459.

Ramage, R., 1972, Carotenoid chemistry. *In:* Newman, A.A. (ed.), Chemistry of Terpenes and Terpenoids, Academic Press, London- New York.

Regnier, F.E., Waller, G.R., and Eisenbraun, E.J., 1967, Studies on the composition of the essential oils of three *Nepeta* species. Phytochem. 6:1281.

Regnier, F.E., Waller, G.R., Eisenbraun, E.J., and Auda, H., 1968, The biosynthesis of methylcyclopentane monoterpenoids--II. Nepetalactone. Phytochem. 7:221.

Reynolds, G., Epstein, W., Terry, W., and Rodriguez, E., 1980, A potent contact allergen of *Phacelia (Hydrophyllaceae)*, Contact Dermatitis 6:272.

Reynolds, G., and Rodriguez, E., 1979, Geranylhydroquinone: A contact allergen from trichomes of *Phacelia crenulata*. Phytochem. 18:1567.

Reynolds, G., and Rodriguez, E., 1981a, Prenylated phenols that cause contact dermatitis from trichomes of *Phacelia ixodes*, Planta Med. (in press).

Reynolds, G., and Rodriguez, E., 1981b, Prenylated hydroquinones: Contact allergens from trichomes of *Phacelia minor* and *P. parryi*, Phytochem. (in press).

Rhoades, D.F., 1977, The antiherbivore chemistry of *Larrea*, Ch. 6 *In:* Mabry, T.J., Hunziker, J.H., and Difeo, Jr., D.R. (eds.), "Creosote Bush, Biology and Chemistry of *Larrea* in New World Deserts", Dowden, Hutchinson & Ross, Stroudsburg.

Ribereau-Gayon, P., 1972, "Plant Phenolics", p. 159. Oliver and Boyd, Edinburgh.

Rice, E.L., 1974, Allelopathy. Academic Press, New York-London.

Roberts, D.L., and Rowland, R.L., 1962, Macrocyclic diterpenes. α- and β-4,8, 13-duvatriene-1,3-diols from tobacco. J. Org. Chem. 27:3989.

Rodriguez, E., Dillon, M.O., Mabry, T.J., Mitchell, J.C., and Towers, G.H.N., 1976a, Dermatologically active sesquiterpene lactones in trichomes of *Parthenium hysterophorus* L. *(Compositae)*. Experientia 32:236.

Rodriguez, E., Towers, G.H.N., and Mitchell, J.C., 1976b, Biological activities of sesquiterpene lactones. Phytochem. 15:1573.

Rodriguez, E., and Levin, D.A., 1976, Biochemical parallelisms of repellents and attractants in higher plants and arthropods. *In:* Wallace, J.W., and Mansell, R.L. (eds.), Biochemical Interaction Between Plants and Insects. Recent Advances in Phytochemistry 10:214.

Rodriguez, E., Epstein, W.L., and Mitchell, J.C., 1977, The role of sesquiterpene lactones in contact hypersensitivity to some North and South American species of feverfew *Parthenium compositae*. Contact Derm. 3:155.

Rodriguez, E., Sanchez, B., Grieco, P.A., Majetich, G., and Oguri, T., 1979, Gerin, a eudesmane methyl ester in trichome exudates of *Geraea viscida*. Phytochem. 18:1741.

Rudali, P.G., and Menetrier, L., 1967, Action del al geranyl-hydroquinone sur differents cancers spontanes et provoques chez les souris, Therapie XXII: 895.

Ruedi, P., and Eugster, C.H., 1971, Struktur von Coleone C, einem neuen Blattfarbstoff aus *Coleus aquaticus* Curcke, Helv. Chim. Acta 54:1606.

Schildknecht, H., Bayer, I., and Schmidt, H., 1967, Struktur des Primelgiftstoffs, Z. Naturforschg. 22b:36.

Schnepf, E., 1969, "Sekretion und Excretion bei Pflanzen", Springer Verlag, Wien.

Schnepf, E., 1974, Gland cells, Ch. 9 In: Robards, A.W. (ed.), "Dynamic Aspects of Plant Ultrastructure", McGraw-Hill, New York.

Scholl, J.P., 1976, Volatile compounds of *Artemisia*, section *Tridentatae* (sagebrush): A study in plant utilization by mule deer. MS Thesis, Univ. Montana.

Schuck, H.J., 1977, Die wirkung von monoterpenen auf das mycelwachstum von *Fomes annosus* (Fr.) Cooke. Eur. J. For. Path. 7:374.

Self, L.S., Guthrie, F.E., and Hedgson, E., 1964, Adaptation of tobacco hornworms to the ingestion of nicotine. J. Insect Physiol. 10:907.

Shade, R.E., Doskocil, M.J., and Maxon, N.P., 1979, Potato leafhopper resistance in glandular alfalfa species, Crop Sci. 19:287.

Shade, R.E., Thompson, T.E., and Campbell, W.R., 1975, Alfalfa weevil larval resistance mechanism detected in Medicago. J. Econ. Entomol. 68:399.

Slama, K., 1978, The principles of antihormone action in insects. Acta Entomol. Bohemoslov. 75:65.

Stahl, E., and Kunde, R., 1973, Neue Inhalsstoffe aus dem ätherischen Ol von *Cannabis sativa*. Tetrahedron Lett. 30:2841.

Stevens, R., 1967, The chemistry of hop constituents. Chem. Rev. 67:19.

Sticher, V.O., and Flück, H., 1968, Die Zusammensetzung von genuinen, extrahierten und destillierten ätherischen Ölen einiger *Mentha*-Arten. Pharm. Acta Helv. 43:411.

Stoner, A.K., 1970, Selecting tomatoes resistant to spider mites, J. Amer. Soc. Hort. Sci. 95:78.

Stoner, A.K., Frank, J.A., and Gentile, A.G., 1968, The
relationship of glandular hairs on tomatoes to spider
mite resistance. J. Amer. Soc. Hort. Sci. 93:532.

Stubblebine, W.H., and Langenheim, J.H., 1977, Effects of *Hymenaea
courbaril* leaf resin on the generalist herbivore *Spodoptera
exigua* (beet armyworm). J. Chem. Ecol. 3:633.

Suga, T., Iwata, N., and Asakawa, Y., 1972, Chemical constituents
of the male flower of *Alnus pendula (Betulaceae)*, Bull.
Chem. Soc. Japan 45:2058.

Thompson, W.W., Platt-Aloia, K., and Koller, D., 1979, Ultrastructure and development of the trichomes of *Larrea
tridentata* (Creosote bush), Bot. Gaz. 140:249.

Thomson, R.H., 1971, "Naturally occurring quinones" 2nd edition,
Academic Press, New York.

Thurston, R., 1970, Toxicity of trichome exudates of *Nicotiana*
and *Petunia* species tobacco hornworm larvae. J. Econ.
Entomol. 63:272.

Thurston, R., Smith, W.T., and Cooper, B.P., 1966, Alkaloid
secretion by trichomes of *Nicotiana* species and resistance
to aphids. Ent. Exp. Appl. 9:428.

Tingey, W.M., and Gibson, R.W., 1978, Feeding and mobility of the
potato leafhopper impaired by glandular trichomes of *Solanum
berthaultii* and *S. polyadenium*, J. Econ. Entomol. 71:856.

Torrance, S.J., Wiedhopf, R.M., and Cole, J.R., 1975, Ambrosin,
tumor inhibitory agent from *Hymenoclea salsola (Asteraceae)*.
J. Pharm. Sci. 64:887.

Turner, C.E., Elsohly, M.A., and Boeren, E.G., 1980, Constituents
of *Cannabis sativa* L. XVII. A review of the natural
constituents. J. Nat. Prod. 43:169.

Turner, J.C., Hemphill, J.K., and Mahlberg, P.G., 1977, Gland
distribution and cannabinoid content in clones of *Cannabis
sativa* L. Amer. J. Bot. 64:687.

Turner, J.C., Hemphill, J.K., and Mahlberg, P.G., 1978,
Quantitative determination of cannabinoids in individual
glandular trichomes of *Cannabis sativa* L. *(Cannabaceae)*.
Amer. J. Bot. 65:1103.

Tyson, B.J., Dement, W.A., and Mooney, H.A., 1974, Volatilisation
of terpenes from *Salvia mellifera*. Nature 252:119.

Ulubelen, A., Miski, M., Neuman, P., and Mabry, T.J., 1979,
Flavonoids of *Salvia tomemtosa (Labiatae)*, J. Nat. Prod.
42:261.

Uphof, J.C. Th., 1962, "Plant Hairs", Gebrüder Borntraeger,
Berlin.

Vanhaelen-Fastre, R., 1972, Activites antibiotique et
cytotoxique de la cnicine, isolee de *Cnicus benedictus*
L. J. Pharm. Belg. 27:683.

Vanhaelen-Fastre, R., and Vanhaelen, M., 1976, Activite
antibiotique et cytotoxique de la cnicine et de ses produits
d'hydrolyse. Planta Medica 29:179.

Vasechko, G.I., 1978, Host selection by some bark beetles (Col. *Scolytidae*) I. Studies of primary attraction with chemical stimuli. Z. Angew. Entomol. 85:66.

Vrkoc, J., Krecek, J., and Hrdy, I., 1978, Monoterpenic alarm pheromones in two *Nasutitermes* species. Acta Entomol. Bohemoslov. 75:1.

Weatherston, J., 1967, The chemistry of arthropod defensive substances. Quart. Rev. 21:287.

Werner, R.A., 1972, Aggregation behavior of the beetle *Ips grandicollis* in response to host-produced attractants. J. Insect Physiol. 18:423.

Widen, C.J., and Britton, D.M., 1971, A chromatographic study of *Dryopteris filix-mas* and related taxa in North American, Can. J. Bot. 49:1589.

Willinsky, M.D., 1973, Analytical aspects of *Cannabis* chemistry. *In:* Mechoulam, R. (ed.), Marijuana: Chemistry, Pharmacology, Metabolism and Clinical Effects, 137-165. Academic Press, New York-London.

LIST OF PARTICIPANTS

H.-Dietmar Behnke, Zellenlehre Universität Heidelberg, Im
     Neuenheimer Feld 230 D-6900 Heidelberg, Fed. Rep. of Germany.

Rodney Croteau, Institute of Biological Chemistry and
     Biochemistry/Biophysics Program, Washington State University,
     Pullman, WA 99164 USA.

James Ehleringer, Department of Biology, University of Utah, Salt
     Lake City, UT 84112 USA.

Charles T. Hammond, Department of Biology, Saint Meinrad College,
     St. Meinrad, IN 47577 USA.

Patrick L. Healey, Department of Developmental and Cell Biology,
     University of California, Irvine, CA 92717 USA.

John K. Hemphill, Department of Biology, Indiana University,
     Bloomington, IN 47401 USA.

Yolande Heslop-Harrison, Welsh Plant Breeding Station, Plas
     Gogerddan, nr Aberystwyth SY23 3EB, UK (manuscript not
     available).

Mark A. Johnson, Institute of Biological Chemistry and
     Biochemistry/Biophysics Program, Washington State
     University, Pullman, WA 99164 USA.

Rick G. Kelsey, Woods Chemistry Laboratory, University of Montana,
     Missoula, MT 59823 USA.

Paul G. Mahlberg, Department of Biology, Indiana University,
     Bloomington, IN 47401 USA.

Robert L. Peterson, Department of Botany and Genetics, University
     of Guelph, Guelph, Ontario, Canada N1G 2W1.

Gary Reynolds, Phytochemical Laboratory, Department of
    Developmental and Cell Biology, University of California,
    Irvine, CA 92717 USA.

Eloy Rodriguez, Phytochemical Laboratory, Department of Ecology
    and Evolutionary Biology and Department of Developmental and
    Cell Biology, University of California, Irvine, CA 92717 USA.

William W. Thomson, Department of Botany and Plant Sciences,
    University of California, Riverside, CA 92521 USA.

E. Laurence Thurston, Electron Microscopy Center, Texas A & M
    University, College Station, TX 77843 USA (manuscript not
    available).

Jocelyn C. Turner, Department of Biology, Indiana University,
    Bloomington, IN 47401 USA.

Janet Vermeer, Department of Biology, Carleton University,
    Ottawa, Ontario, Canada K1S 5B6.

Eckhard Wollenweber, Institut für Botanik der Technischen
    Hochschule, Schnittspahnstraße 3, 6100 Darmstadt, Fed. Rep.
    of Germany

INDEX

Abietane, 159
Abronia villosa, 121, 122
Absinthin, 194, 201
Achillea, 73, 84-87
   A. millefolium, 72, 88
Achyranthes aspera L., 217
Acid polysaccharides, 1
Acumen, 6, 9
Adhumulone, 196, 203, 212
Adlupulone, 196, 203
Aesculus, 53
   A. Carnea, 55, 58
   A. hippocastanum, 55, 58, 85
   A. indica, 55, 58
   A. turbinata, 55, 58
Ageratriol, 166, 167
Agerol, 166, 167
   diepoxide, 166, 167
Alkaloids, 228
Alkylated phenolics, 187, 224
Alnus, 55, 221, 223
   A. glutinosa, 59
   A. pendula, 224
   A. sieboldiana, 224
Alnustone, 223, 224
Ambrosia cumanensis, 217
Ambrosin, 194, 201, 209, 211
Amitermes vitosus, 206
Anabasine, 228
Anabsinthin, 194, 201
Andrographis paniculata, 153
Ants, 205
Apigenin, 55, 56, 67
   4'-methyl ether, 56, 64
   7-methyl ether, 64, 66
   7,4'-dimethyl ether, 66

Apigenin (continued)
   7,3'-dimethyl ether, 66
Apocynaceae, 228
Apteria cordifolia, 75
Arabidopsis sp, 4
Arabinose, 10
Arctostaphylos, 117
Aridity gradient, 115
   and pubescence, 115-119
Artabsin, 194, 201
Artemeseole, 192, 199
Artemisia, 67, 117
   A. absinthium L., 194
   A. nova, 190, 191, 194
   A. nova Nelson, 190, 214, 218
   A. tridentata ssp. *vaseyana*, 190, 192
Asteraceae, 67, 117
Atisirane, 159
Atriplex, 95-98, 105, 106, 127
Avena fatua L., 214, 216
Avicennia, 103, 104

Bacillus,
   B. cereus, 212, 220
   B. megaterium, 212
   B. subtilis, 212, 220
Bacteria, 1, 9, 10, 14, 18, 19, 75
   and trichome initiation, 9
   N-fixing, 9
   parasitic, 19
   symbiotic, 19
Bacterial symbiosis, 9
Barbacenia purpurea, 65
Bark beetles, 205

Battus polydamas, 204, 206
α-bergamotene, 139, 154
  trans-α-bergamotene, 193, 200
Betula, 55
  B. nigra, 55
Betulaceae, 53, 55, 57, 86, 221, 224
Betulic acid, 196, 203
Beyer-15-en-19-ol, 198
Beyer-15-en-3-one, 6,17-diol, 196, 202
Beyer-15-en-3-one, 6-OAc-17-ol, 195, 202
Beyer-15-en-3-one-17-oicacio, 19-OAc, 195, 202
Beyerane, 159
Beyeren-19-ol, 166, 167
d-beyerene, 160, 161, 166, 167
Beyeria, 67, 189
  B. leschenaultii, 166, 167, 195, 196
  B. viscosa, 218
Beyerol, 166, 167, 198
Bicyclogermacrene, 139
Bidens pilosa L., 217
Bisabolene, 151, 153, 155, 156
  β-bisabolene, 193, 200
  α-bisabolol, 139, 145
Bladder cell,
  of salt glands, 96-99, 101, 106
Boletus variegatus, 209, 211
Bombyx mori, 204, 206
Boraginaceae, 227
Boranes, 137
Bordetella bronchiseptica, 211
Borneol, 198
  d-borneol, 138, 147, 149, 154, 168
Bornyl pyrophosphate, 147-149
Botrytis cinerea, 212, 220
Brassicaceae, 4, 5
Brickellia, 67, 117
Bromus rigidus Mex, 214, 216
Brucella abortus, 211
Bryonia dioica, 72, 73
Bulnesene, 193, 200

Cacti, 120, 126

Cadalene, 139
γ-cadinene, 140, 198
δ-cadinene, 193, 200, 207
Caffeic acid, 225
Cajaninae, 5
Calcite (calcium carbonate), 74
Camphene, 154, 190, 199, 204, 206, 209, 210, 214, 216
Campherenol, 154
Campherenone, 139
Camphor, 190, 199, 207, 210, 214, 216, 220
  d-camphor, 138, 147, 149, 168
  l-camphor, 149
Candida albicans, 207, 210
Cannabaceae, 9
Cannabichromene, 197, 203
Cannabidiol, 37, 39, 43, 197, 203, 212
Cannabigerol, 197, 203, 212
Cannabinoids, 23, 24, 30, 36, 39, 40, 48, 49, 75, 218
  contents, 23, 24, 38
  in plant organs, 37-40
  in individual glands, 39-43
  and development, 37-39
  and cell fractionation, 48
  synthesis, 23, 44
  terpenophenolic, 48
Cannabinol, 197, 203
Cannabis, 23-25, 31, 35, 37, 43, 44, 46, 74, 75, 87, 88
  C. sativa, 6, 9, 23, 48, 49, 87, 189, 191-194, 197, 207, 209, 218
  gland description, 25-30
  gland development, 30-33
  gland populations (density) 34-37
  gland ultrastructure, 43-48
Capitate hairs, 2
Carbohydrates, 82, 83, 96, 107, 109
Carotenoids, 189
Carum carvi, 153
Carvacrol, 165, 198
Caryophyllene, 204, 207, 214, 217
  β-caryophyllene, 193, 200, 206

INDEX

Caryophyllene (continued)
  β-caryophyllene (continued)
    effect on insects, 204
Casbene, 157, 158
Cassane, 159
Castor bean, 157, 161
Ceanothus, 117
Cecropiaceae, 6
α-cedrene, 140
Cell wall, 86, 87
  cutinized, 86
Cembrene A, 158
Cembrenetriol, 195, 202, 207
Centrosema, 6
Ceratocystis, 209
  C. minor, 210, 211
Ceroptin, 54
Chalcones, 64 (see specific compounds)
Chalk glands, 74
Chamomile, 84
Chenopodiaceae, 117
Chieilanthes, 61, 65
Chrysanthemum, 73, 78, 81, 87, 88
  C. morifolium cv *dramatic*, 71, 77, 83, 85
  C. parthenium Bernh., 194, 209
Chrysin, 56, 66, 67
  7-methyl ether, 56
Chrysothamnus, 67
Cineole, 190, 199, 204, 216, 220
  1,8-cineole, 138
Citronellol, 138
Citrus, 150, 168
Cleome spinosa, 193
Clitoria, 6
Clitoriopsis, 6
Cnicin, 194, 201, 211
Cnicus benedictus L., 194
Cohumulone, 197, 203, 213
Coleone, 223
  Coleone C., 225
Coleus, 223
  C. aquaticus, 225
Colleters, 2
Colletotrichum acutatum, 212
Colupulone, 197, 203, 213
Condalia, 117

Copaene, 198
Copalyl pyrophosphate, 160, 161
Cordia alliodora, 227
Corynebacterium, 207
Cruciferae, 1
Crusea calocephala D.C., 217
Cryptococcus albidus, 212
Crystals, 74
  Calcium oxalate, 73
    intra nuclear, 73, 75
    protein, 74
Cubitene, 207, 208
Cubitermes umbratus, 207, 208
Cucumia sativus, 72, 214, 216, 217
Cumambrin,
  A, 194, 201, 214, 217, 220
  B, 194, 201, 214, 217, 220
Cuparene, 139
Curcumene, 191, 200
  α-curcumene, 139
  β-curcumene, 139
  γ-curcemene, 154
Cuticle, 87, 88
p-cymene, 138, 164, 165
p-cymene-8-ol, 165
p-cymol, 198
Cynondon, 100
Cypripedium, 225
Cystolith trichomes, 74
Cytoplasm, 10, 44, 46, 48, 82, 96, 98

Decipiane triol, 196, 202
Dendroctonus micans, 205
Dictyosomes, 10, 14, 26, 44, 82, 86, 99
Didymocarpus pedicellatus, 65
Dihydrochalcones, 64, 65
Dihydronepetalactone, 198
6β-17-dihydroxybeyer-15-en-3-one, 166
17,19-dihydroxybeyer-15-en-3-one, 166
Dihydroxyserrulatic acid, 195, 202
Digestive glands, 71, 85
Dimethylallyl pyrophosphate (see DMAPP)

Dionysia, 55
Dioscorea,
  D. macroura, 1, 6, 9, 10, 13, 14, 17, 18
  D. sansibariensis, 6
Dioscoreaceae, 6
Diphenylheptanoid compounds, 224
Diterpenes, 133, 134, 187, 189, 195, 198, 202, 207, 208, 218, 219
  bicyclic, 195, 202
  biological significance, 172
  biosynthesis, 143
    cyclization, 155
    secondary transformation, 162
  catabolism, 169
  in glandular trichomes, 195, 196, 202
  monocyclic, 195, 202
  site of synthesis, 135
  tricyclic, 195
Diterpenoid quinones, 225
DMAPP (dimethylallyl pyrophosphate), 134, 143, 145
Dopaldehyde, 225
Dryopteris, 221, 223, 224
  D. marginalis, 224
α-4,8,13-duvatriene-1,3-diol, 195, 202
β-4,8,13-duvatriene-1,3-diol, 195, 202

Eccrine secretion, 96, 99, 105, 108
Echinocystic lobata, 72
Eleusine coracana, 214, 216, 217
Elodea, 82, 86
Empoasca fabae, 224
Encelia, 67, 117, 118, 120, 125, 129
  E. actonii, 120
  E. californica, 118, 119, 128
  E. farinosa, 117-120, 126-130
  E. frutescens, 120
  E. palmeri, 120
  E. virginensis, 117, 118, 120, 125
Enceliopsis, 117, 118

Enceliopsis (continued)
  E. argophylla, 117, 118, 123
Endoplasmic reticulum, 26, 44, 85, 86, 96, 102, 105, 107, 109
  rough ER, 82, 99, 107, 108
  smooth ER, 82, 107, 108
Enmcin, 166
Enzymes, 85, 86
  histochemical test for, 85
Eperuane-8β,15-diol, 195, 202
Epinepetalactone, 191
α-1,7a-epinepetalactone, 199
Epoxycembradienol, 195, 202, 207
Eremolactone, 196, 202
Eremophila, 73, 87, 189
  E. clarkei F. Mvell, 195, 207
  E. decipiens Ostenf, 196
  E. fraseri, 74, 75, 196, 218
  E. georgei Diels, 195, 207
  E. serrulata F. Muell, 195
Eremophilane, 156
Ericaceae, 117
Ericameria, 67
Eriogonum, 117
Erythrina, 5
Escallonia, 65, 66
Escherichia coli, 207, 210
Essential oils, 71, 83, 84
  histochemical test for, 84
Eudesmane, 151, 156
Eupatorium, 67
Euphorbiaceae, 67
Exocytosis, 107

Fabaceae, 5, 6, 228
β-farnesene, 139, 154, 191, 200
Farnesyl hydroquinone, 227
Farnesyl pyrophosphate (see FPP)
Fats, 85
Fatty acids, 224
l-α-fenchol, 138, 149
d-fenchone, 138
Festuca megalura Nutt, 214, 216
Flavonoids, 71, 85, 107, 218, 219
  aglycones, 83, 85, 224
  glycosides, 222
  methylated, 187, 222

INDEX

Flavones
  3'4'-dihydroxyflavone, 55, 60
  5,2'-dihydroxyflavone, 55, 60
  5,8-dihydroxyflavone, 55, 60
  5-hydroxyflavone, 55, 60
  2'-hydroxyflavone, 55, 60
  6-methoxyflavone, 55, 60
  5,8,2'-trihydroxyflavone, 55, 60
Flavanone (see specific compounds)
Flies, 205
Flourensia, 67
Foeniculum vulgare, 149, 168
Fomes annosus, 209-211
Formica rufa, 205
FPP (farnesyl pyrophosphate), 134, 143-145, 151, 152, 155
Frankenia, 86, 102, 105
Franseria, 117

Galactose, 10
Geraea, 118
  G. canescens, 117, 118
  G. viscida, 191
Geraniol, 138
Geranyl acetate, 198
Geranylbenzoquinone, 225, 227
Geranylgeranylhydroquinone, 227
Geranylgeranyl pyrophosphate (see GGPP)
Geranyl isobutyrate, 198
Geranyl proprionate, 198
Geranyl pyrophosphate (see GPP)
Gerea, 67
Gerin, 191, 200
Germacrene,
  C, 139, 153
  D, 139, 153
Gesneriaceae, 65
GGPP (geranylgeranyl pyrophosphate), 134, 143-145, 157-161
Gibberellins, 214
Giberella fujikori, 152
Gland initiation in *Cannabis*, 35, 37
Golgi vesicles/bodies, 10, 82, 85, 86, 107

GPP (geranyl pyrophosphate),
GPP (continued)
  134, 143-147, 149, 150, 152, 163, 165
Granulocrine secretion, 96
Grewia, 75, 86
Guaiene, 207, 211
  α-guaiene, 191, 200
Gymnogramoid ferns, 53, 61

Haplopappus, 67
Hedycaryol, 139
Heliothis zea, 224
Holocrine secretion, 96, 99
Hooked hair, 5, 6
Hops (see *Humulus lupulus*)
Hordeum leporinum Link, 214, 216
Humulane, 151, 156
  β-humulene, 191, 200
Humulene epoxide, 198
Humulenol, 198
Humulone, 197, 213
  α-humulone, 207
  β-humulone, 207
Humulus, 87
  H. lupus, 6, 9, 87
  H. lupulus L., 189-194, 196, 197, 204, 209
  H. spinosa L., 190
Hydrophyllaceae, 67, 187, 221, 225
2-(1-oxofarnesyl)-hydroquinone, 227
6-hydroxy kaempferol, 55
  3,6-dimethyl ether, 57
  6,4'-dimethyl ether, 57
  6,7-dimethyl ether, 57
  3,6,4'-trimethyl ether, 57
  6,7,4'-trimethyl ether, 57
17-hydroxy-19-norbeyer-15-en-3-one, 198
Hylvrogops glabratus, 205
Hymenaea covrbaril, 207, 215
Hypera postica, 224

Indole alkylamines, 228
Indumentum, 1-3, 5, 114, 120, 127
IPP (isopentenyl pyrophosphate), 134, 143, 145, 189

Ips grandicollis, 205
I. typographus, 205
Irradoids, 204
l-isoborneol, 138
Isocaryophyllene, 191, 200
d-isomenthol, 163
Isomenthone, 198
Isopentenyl pyrophosphate (see IPP)
l-3-isothujone, 138, 168
Isopimara-9(11), 15-diene 3-β-19-diol, 196, 202
Isoprenoids, 187

Kadsura japonica, 153
Kaemferol, 55, 56
  3,4'-dimethyl ether, 57, 66
  3,7-dimethyl ether, 57, 66
  7,4'-dimethyl ether, 65, 66
  3-methyl ether, 56, 66
  4-methyl ether, 56
  7-methyl ether, 56, 65, 66
  3,7,4'-trimethyl ether, 57
Kaur-16-en-19-oic acid, 166, 167
Kaur-16-ene-15-one, 166
Kaurane, 145, 159, 160, 161, 167
d-kaurene, 140, 145, 166
Kolauenol, 140
Kochia, 117

Labda-8(17),13-dien-15-yl pyrophosphate, 159
Labiatae, 189, 204, 225, 227
Lactobacillis, 212, 213
  L. pastorianus, 212, 213
  L. plantarum, 212, 213
Lamiaceae, 65, 67, 117
Lanceol, 139
Lantana, 87
Larrea, 65, 187, 221-223
  L. tridentata, 220
Leaf
  absorptance, 113, 115, 116, 118-120, 126-130
    spectrum, 115, 118, 119
  pubscences, 1, 113-115, 117, 118, 126, 129, 130
  reflectance, 113-115, 118, 120, 129, 130

Leguminosae, 1
Lenzites saepiaria, 209-211
Lesquerella, 5, 72
Lignans, 187, 222
Limonene, 138, 140, 168, 198
Limonium, 102
Linalool, 138, 154, 198
Linalylacetate, 198
Linalylproprionate, 198
Linaloyl pyrophosphate (see LPP)
Linalylvalerianate, 198
Lipid, 11, 85, 89, 107-109
  secreting glands, 95
Lipophilic substances, 83, 85, 96, 108
  histochemical test for, 83
Lithocysts, 9
Longifolene, 139, 155, 156, 191, 200, 209, 211
LPP (linaloyl pyrophosphate), 134, 146
Lupulone, 197, 213
Luteolin, 65
  3-methyl ether, 65
Lycopersicon, 222
  L. esculentum, 74

Madina sativa Molina, 214, 216
Majorana hortensis, 135, 192
Malvaceae, 117
Malvastrum rotundifolium, 121, 122
Mangroves, 102, 103
d-manool, 140
Marah macrocarpus, 161, 162
Margaspidin, 223, 224
Marihuana (see *Cannabis sativa*)
Marrubium vulgare, 169
Medicago, 85, 224
  M. disciformis, 224
  M. sativa, 224
Mentha
  M. aquatica, 191, 193
  M. pierita, 83, 163, 164, 190, 191, 204, 218
p-menthanes, 137
Menthene, 198
Menthofuran, 191, 199

INDEX

Menthol, 190, 199, 204, 206, 207, 210
  effects on insects, 204
  l-menthol, 163, 170, 171
Menthone, 190, 199, 204, 206
  d-menthone, 163
  l-menthone, 163, 170, 171
Menthylacetate, 170, 171, 190, 198, 199
Menthylgeranate, 198
Mercurialis, 75
Methyl eugenol, 84
Mevalonic acid (see MVA)
Micrococcus lysodeikticus, 212, 220
Micro-hairs, 1
Micropappillate surface, 4-6, 9
Microvacuoles, 105
Mimulus tilingii, 82
Mitochondria, 26, 44, 82, 84, 101, 102
Mixed terpenoids, 189, 196, 203, 209, 212
  effect on bacteria, 212, 213
  in glandular trichomes, 196, 197, 203
Mohavea breviflora, 121, 122
Monarda fistulosa, 84
Monterpenes, 133, 134, 187-190, 198, 199, 206, 207, 209, 216, 218
  acyclic, 190
  bicyclic, 190, 195
  biological significance, 172
  biosynthesis, 141
    cyclization, 146
    secondary transformations, 162
  catabolism, 169
  effects on bacteria and fungi, 210, 211
  in glandular trichomes, 190-192, 199
  irregular, 192
  monocyclic, 190, 195
  phytotoxic effects, 216
Moraceae, 6
Mucilage hair, 82, 85, 87, 95
Mucopolysaccharide, 10, 13, 14,

Mucopolysaccharide (continued) 18, 19
Mucuna pruriens, 228
Musca domestica, 204-206
α-muurolene, 139
MVA (mevalonic acid), 134-136, 142, 155, 169
Mycobacterium, 212, 220
Myrcene, 138, 154, 190, 199, 204, 206, 209, 210
Myricetin 7,3'4'-trimethyl ether, 58
Myrmicaria eumenoides, 205

Narigenin
  7,4'-dimethyl ether, 57
  4'-methyl ether, 57
  7-methyl ether, 57, 67
Nasutitermes
  N. costalis, 205
  N. rippertii, 205, 207
NDGA (nordihydrguaiaretic acid), 221-223
Nectary trichomes, 71
Neocembrene-A, 207, 208
d-neoisomenthol, 163
l-3-neoisothujanol, 138, 168
Neomenthol, 198
  d-neomenthol, 163, 170, 171
Nepeta cataria L., 191, 193, 204, 218
Nepetalactone, 191, 199, 204, 206
  effect on insects, 204
Nerol, 138, 190, 199, 207, 210
Nerolidiol, 198
Nerolidol, 139, 154
Neryl pyrophosphate (see NPP)
Nesquehonite (magnesium carbonate), 74
Newcastelia, 189
  N. viscida, 85, 196, 209, 219
Nicotene, 228
Nicotiana, 74, 88, 228
  N. tabacum, 195, 228
Nocardia, 220
Nordihydroguaiaretic acid (see NDGA)
Nornicotene, 228
Notholaena, 61, 65

Notholaena (continued)
  N. candida var. *coplandii*, 63
  N. grayi, 65
  N. greggii, 65
  N. sulphurea, 65
NNP (neryl pyrophosphate), 134, 144, 146, 149, 150, 165
Nucleus, 26, 72, 73, 82, 101

Ocimene, 198
Ocimum gratissimum, 84
Oleanolic acid, 196, 203, 209, 212
Ophrys orchids, 172
Opium alkaloids, 225
Oridonin, 166
Ostrya, 55

Paniculide β, 153, 156
Papillae, 1-3
Parthenin, 194, 201, 209, 212, 217, 220
Parthenium hysterophorus L., 194, 209, 218
Parthenolide, 194, 201, 209, 212, 220
α-patchoulene, 192, 200
β-patchoulene, 192, 200
γ-patchoulene, 192, 200
Patchouli alcohol, 192, 200
Pelargonium, 78, 81
  P. graveolens, 71, 77, 83
Peltate hairs, 1, 2
  scales, 9
Penneaceae, 65
Peppermint, 168-171
Periandra, 6
Petasin, 155, 156
Petasites hydridus, 155
Petunia, 228
Phacelia, 95, 96, 107-109, 221, 225, 226
  P. calthafolia, 121, 122
  P. crenulata, 225
  P. ixodes, 225
  P. minor, 227
  P. parryi, 227
Pharbitis, 73, 95, 96, 107-109
  P. nil, 75, 83

Phaseoleae, 5
Phaseolus s.l., 6
Phaseolus vulgaris, 6, 214, 217
β-phellandrene, 138, 153
Phenolic, 74, 75, 89, 96, 108, 109
  acid, 224
  aglycones, 222
  alcohols, 224
  constituents, 220, 223, 226
    in trichomes, 221
  esters, 224
  histochemical test for, 74
  stilbenes, 221, 224
Phenylated quinones, 107
Phenylpropanoid derivatives, 224
Phloroglucinol derivatives, 221, 224
d-phyllocladene, 140
Physaria, 72
Pimarane, 159
Pimaric acid, 140
Pinanes, 137
α-pinene, 138, 145, 150, 154, 191, 199, 204-206, 209, 211, 214, 216, 220
  effects on insects, 204, 205
l-β-pinene, 138, 140, 145, 150, 198
Pinobanksinacetate, 56
Pinocembrin, 56, 66
  7-methyl ether, 56
Pinocytosis, 82
Pinus, 155
  P. ponderosa, 209
Piperitenone, 163
Piperitone, 198
Pityrogramma, 53, 54, 64, 65, 85
  P. austroamericana, 62
  P. chrysophylla, 61
  P. lehmanii, 61
  P. triangularis, 54, 61
    var. *maxonii*, 61
    var. *pallida*, 61
    var. *semipallida*, 61
    var. *triangularis*, 61
    var. *viscosa*, 64
Plasmalemma, 10, 13, 14, 83,

Plasmalemma (continued)
    86, 96, 101, 102, 105–109
Plastids/chloroplasts, 44, 46,
    73, 74, 78, 82, 85, 96, 109
Plectranthus, 223
    P. caninus, 225
Plumbago, 85
    P. capensis, 74
Pogonomyremex badius, 204
    P. occidentalis, 205
Pogostemon cablin, 191–193, 219
Polygonaceae, 117
Polyphenol oxidase, 222
Polyploidy, 72–73
Polypodiaceae, 221
Polypodium, 78
    P. vulgare, 71
    P. virginianum, 73, 75, 77,
        83
Polysaccharides, 48, 71, 82, 83,
    87–89
    histochemical test for, 82
Polytrichiales, 82, 85
Polytrichium, 87
Populus, 53–56
    P. balsamifera, 54
    P. deltoides, 56
    P. euramericana, 56
    P. Koreana, 56
    P. Maximov, 56
    P. nigra, 56
    P. Simonii, 56
    P. trichocarpa, 56
Prenylated hydroquinone
    derivatives, 221
Prenylated phenolic compounds,
    225, 226
Primin, 221, 223, 225
Primula, 53–55, 60, 61, 221,
    223, 225
    P. obconica, 224
    P. sinensis, 55
Primulaceae, 55, 187, 221, 224,
    225
Pristiphora erichsonii, 204–206
Pristomyrmex pungens, 204–206
Proteins, 71, 75–82, 89, 96, 107
Psathyrotes ramosissima, 123
Pseudomonas aeruginosa, 211

Psychotria, 87
    P. bacteriophila, 75
Pulegone, 198
    d-pulegone, 163, 170
Pyrrhocoris apterus, 204, 206

Quantum flux, 128, 129
Quercetagetin, 67
Quercetin, 55, 67
    3,3'-dimethyl ether, 56, 57,
        66
    3,7-dimethyl ether, 56, 57
    7,3'-dimethyl ether, 56, 58,
        66
    3'-methyl ether, 58, 65, 66
    7-methyl ether, 56, 66
    3,6,7-trimethyl ether, 65
    3,7,4'-trimethyl ether, 57, 66
    7,3',4-trimethyl ether, 57, 58
Quinones, 187, 224, 225
Quinonoid compounds, 224, 225

Raphanus sativus, 214, 216
Resins, 83, 85
    histochemical test for, 85
    secreting trichomes, 87
Rhamnaceae, 117
Ribosomes, 98, 99, 101, 102, 105
Ricinocarpos muricatus, 195
Ricinus communis, 160, 161
Rimuene, 140
Root hairs, 1
Rosaceae, 86
Rosane, 159
Rizopogon roseolus, 209, 211
Rubiaceae, 9, 67, 86

Sabinene, 192, 199
    hydrate, 154
        cis-sabinene hydrate, 192,
            199
        trans-sabinene hydrate, 198
Saccharomyces carlsbergensis,
    209, 212
Salt glands, 71, 86, 95, 96,
    100–106, 108, 127
Saltera sarcocolla, 65
Saliva, 115–118, 120, 172
    S. apiana, 115, 116, 214

Saliva (continued)
  S. glutinosa, 65
  S. luecophylla, 115, 116,
    124, 214
  S. mellifera, 115, 116,
    214
  S. officinalis, 149, 153
  S. sclarea, 169
d-sandaracopimara-8(14),16-
  diene, 157, 160, 161
β-santalane, 154, 198
Sativene, 139
Scales, 1-3, 5, 10
  glandular (squamule), 19, 82,
    86
Scizophyllum commune, 209-211
Sclareol, 140, 169
Sclerospora graminicola, 209,
  212
Scutellarein, 67
  6,4'-dimethyl ether, 57
  6,7,4'-trimethyl ether, 57
Saxifragaceae, 65, 66
3,4 secobeyerene acid, 116, 196,
  202
Selina-3,7(11)-diene, 192, 200
Selina-4(14),7(11)-diene(34),
  192, 200
Selinane, 151
α-selinene, 193, 200, 207
β-selinene, 194, 200, 204, 206,
  207
Serotonin, 228
Sesquiborane, 151
Sesquisabinene, 154
Sesquiterpene(s), 133, 134, 187,
  189, 193, 198, 200, 204,
  206, 207, 209, 211, 214,
  215, 217-220, 228
  biological significance, 172
  biosynthesis, 143
    cyclization, 150
    secondary transformations,
      162
  catabolism, 169
  effects on bacteria and
    fungi, 211, 212
  hydrocarbons, 193, 215
  in glandular trichomes, 193,

Sesquiterpene(s) (continued)
  in glandular trichomes,
    (continued)
    194, 200, 201
  lactones, 194, 201, 207, 209,
    214, 215, 218-220, 228
  phytotoxic effects, 217
  sites of synthesis, 134
Sesquithujene, 139
Solanaceae, 221, 228
Solanum, 2, 84
  S. benthaultii, 222
  S. nigrum, 73
  S. poladenium, 221
  S. polyadenium, 222
  S. tarijense, 222
  S. tuberosum, 74, 222
Spartocytisus filipes, 88
Sphaeralcea, 117
Spodoptera exigua, 207, 215
Squamules (see scales,
  glandular)
Staphylococcus aureus, 207, 210,
  211
Steviol, 166, 167
Stilbenes, 223

Tamarix, 102, 105
Tanacetum vulgare, 168, 169
Tanins, 74, 75, 221
Termites, 205
Terpenes, 83, 84, 189, 214
  terpene pheromones, 204
α-terpinene, 138, 154, 198
γ-terpinene, 138, 151, 153, 154,
  164, 198
Terpinen-4-ol, 138, 169
α-terpineol, 138, 198
Terpenoids, 188-190, 198, 206-
  210, 214, 215, 229
  concentration, 215, 218, 219
  effects on bacteria and fungi,
    207-213
  effects on insects, 206
  hormones, 204, 214, 215
    gibberellins, 214
    abscisic acid, 214
  in glandular trichomes, 190-
    198

INDEX

Terpenoids (continued)
  phytotoxic effects, 214, 216-219
Terpinolene, 138, 163, 164
Tetradymia, 117
Δ^9-tetrahydrocannabionol, 37, 40, 41, 43, 197, 203
Tetranychus, 222
α-thujene, 198
Thujone, 192, 199
  d-3-thijone, 138
Thymol, 84, 138, 164, 165, 198
Thymus vulgaris, 153, 162, 164
Tobacco, 73, 85
Tonoplast, 96, 101, 105
Trachylobane, 159, 160, 161
Trans,trans-farnesol, 139
Trichoderma viride, 210, 211
Trichome-bacteria interaction, 1, 6-19
Trichomes:
  bulbous, 23, 25, 26, 30, 31, 34-37
  capitate, 72, 85, 88
    -sessile, 23, 25, 26, 30, 31, 34-37, 41-43, 75
      antherial, 25, 26, 30, 31, 35
    -stalked, 23, 25, 26, 28, 30-32, 35-37, 39, 41-44, 75
  classification of, 2-6
  in *Cannabis*, 31
  complex (branched), 1-3, 115, 118, 120
    uniseriate, 1, 3
    multiseriate, 1, 3
  dendritic, 1, 3-5
  (gland) density in *Cannabis*, 34
  development in *Cannabis*, 30-31
  distribution in *Cannabis*, 35
  glandular/secretory, 1-3, 5, 9, 10, 13, 14, 17, 23-25, 71-75, 77, 78, 81, 83-89, 95, 96, 107-109, 120
    of *Cannabis sativa* L., 23-51

Trichomes (continued)
  histochemistry, 71-94
  non-glandular, 2, 3, 5, 25, 26, 37, 71, 75, 87, 109, 120
  simple (unbranched), 1-3, 6, 120
    uniseriate, 1, 3, 5, 9, 10, 17-19
    multiseriate, 1, 3
    stellate, 1, 3, 5, 115
  terminology, 2-4
  T-shaped, 1, 5, 88
  ultrastructure, 4-6, 95
    Cannabis, 43-48
Trifluoracetic acid, 10
Trineruitermes bettonianus, 206
2,3,5-triphenyltetrazolium chloride, 85
Triterpenes, 189, 196, 203, 212, 218, 219
  effects on fungi, 212
  in glandular trichomes, 196, 203

Urticales, 1, 6

Vacuoles, 44, 86, 98, 99, 101, 102, 105, 106
Valeriana, 84
Velloziaceae, 65
Verticillol, 158
Vesicular glands, 5
β-vetivone, 139
Viguiera, 117
Vinca herbacea, 228

Waxes, 83, 120
Wigandia kunthii, 227
Wormwood diterpene, 140

Xylose, 10

Zerumbone, 139
Zygophyllaceae, 65, 187, 221